AN ILLUSTRATED GUIDE TO THE
ANIMALS OF
AMERICA

AN ILLUSTRATED GUIDE TO THE
ANIMALS OF
AMERICA

A VISUAL ENCYCLOPEDIA OF AMPHIBIANS, REPTILES AND MAMMALS IN THE
UNITED STATES, CANADA AND SOUTH AMERICA, WITH OVER 350 ILLUSTRATIONS

TOM JACKSON
CONSULTANT: MICHAEL CHINERY

ARMADILLO

This edition is published by Armadillo an imprint of Anness Publishing Ltd
Blaby Road, Wigston, Leicestershire LE18 4SE; info@anness.com

www.annesspublishing.com

Anness Publishing has a new picture agency outlet for images for publishing, promotions or advertising.
Please visit our website www.practicalpictures.com for more information.

Publisher: Joanna Lorenz
Project Editor: Felicity Forster
Copy Editors: Gerard Cheshire, Richard Rosenfeld and Jen Green
Illustrators: Stuart Carter, Jim Channell, John Francis,
Stephen Lings, Alan Male, Shane Marsh and Sarah Smith
Map Illustrator: Anthony Duke
Designer: Nigel Partridge
Production Controller: Mai-Ling Collyer

Illustrations appearing on pages 1–5: page 1 American bison; page 2 American black bear; page 3 long-nosed armadillo;
page 4t emerald tree boa; page 4b night monkey; page 5 (top row, from left to right) squirrel, beaver, cougar; page 5
(bottom row, from left to right) moose, prairie dogs, brown bear

Previously published as part of a larger volume, *The World Encyclopedia of Animals*

Manufacturer: Anness Publishing Ltd, Blaby Road, Wigston, Leicestershire LE18 4SE, England
For Product Tracking go to: www.annesspublishing.com/tracking
Batch: 6700-22320-1127

CONTENTS

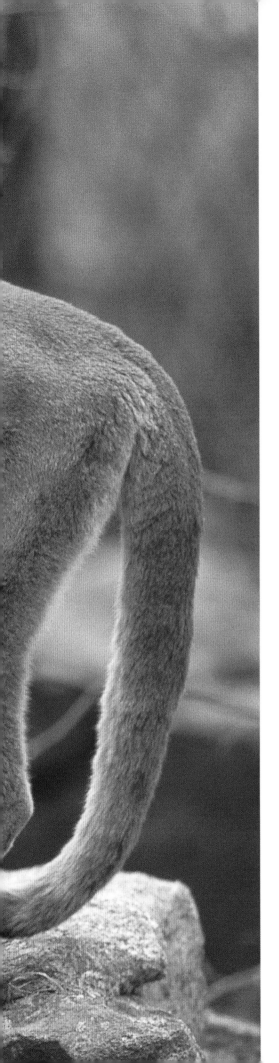

UNDERSTANDING ANIMALS

This book examines amphibians, reptiles and mammals of America, but can only scratch the surface of the fantastic range of life forms packed into the United States, Canada and South America. It concentrates on the vertebrates (the group of animals that have a backbone, which includes the above three groups as well as birds and fish); however, vertebrates form just one of the 31 major animal groups. The huge diversity of life contained in the other groups cannot rival the most familiar vertebrates for size, strength or general popularity. For most people, slugs are not in the same league as cougars, nor can crickets and crabs compete with dolphins and bears. Yet these relatively unpopular animals are capable of amazing feats. For example, squid travel by jet propulsion, mussels change sex as they get older, and ants live in harmony in colonies of millions.

The diversity of life among the vertebrates is mind-boggling. A blue whale is 30m (100ft) long, while the smallest salamander is less than 3cm (1¼ in) long. The rest appear in every shape and size in between. Yet vertebrates are not distinguished by their bodies alone but by the unusual ways in which they use them. For example, a lizard will shed its tail if a hungry predator clamps its jaws around it, and while the tail-less reptile escapes, its attacker is kept occupied by the still-wriggling appendage. The paradoxical frog is another surprise. While every other vertebrate grows as it gets older, this amphibian actually shrinks so that the tadpoles are larger than the adults. And hunting bats, which can fly thanks to the skin stretched over their elongated hand bones, so creating wings, manage to locate their prey in the dark by using sonar. They bark out high-pitched calls that bounce off surrounding objects, and the time taken for the echo to return signifies the size and location of whatever is out there. These are just three examples of the extraordinary diversity found in the vertebrate kingdom.

Left: The cougar – also known as the puma, panther or mountain lion – is the largest of the small cats in America (other American small cats include the margay, ocelot, pampas cat, bobcat and jaguarundi). Cougars live alone, patrolling large territories and hunting mule deer and moose. They are extremely agile animals, being able to leap over 5m (16.5ft) into the air.

EVOLUTION

Animals and other forms of life did not just suddenly appear on the Earth. They evolved over billions of years into countless different forms. The mechanism by which they evolved is called natural selection. The process of natural selection was first proposed by British naturalist Charles Darwin.

Many biologists estimate that there are approximately 30 million species on Earth, but to date only about two million have been discovered and recorded by scientists. So where are the rest? They live in a staggering array of habitats, from the waters of the deep oceans where sperm whales live to the deserts of Mexico, inhabited by the powerful, poisonous, gila monster lizard. The problems faced by animals in these and other habitats on Earth are very different, and so life has evolved in great variety. Each animal needs a body that can cope with its own environment.

Past evidence

At the turn of the 19th century, geologists began to realize that the world was extremely old. They used animal fossils – usually the hard remains, such as shells and bones, which are preserved in stone – to measure the age of the exposed layers of rock found in cliffs and canyons. Today we accept that the Earth is about 4.5 billion years old, but in the early 1800s the idea that

the world was unimaginably old began to change people's ideas about the origins of life completely.

In addition, naturalists had always known that there was a fantastic variety of animals, but now they realized that many could be grouped into families, as if they were related. By the middle of 19th century, two British biologists had independently formulated an idea that would change the way that people saw themselves and the natural world forever. Charles Darwin and Alfred Wallace thought that the world's different animal species had gradually evolved from extinct relatives, like the ones preserved as fossils.

Darwin was the first to publish his ideas, in 1859. He had formulated them while touring South America where he studied the differences between varieties of finches and giant tortoises on the Galápagos Islands in the Pacific Ocean. Wallace came up with similar ideas about the same time, when studying different animals on the islands of South-east Asia and New Guinea.

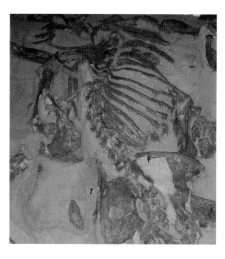

Above: Scientists know about extinct animals from studying fossils such as these mammoth bones. Fossils are the remains of dead plants or animals that have been turned to stone by natural processes over millions of years.

Survival of the fittest

Both came up with the same idea – natural selection. As breeders had known for generations, animals pass on their characteristics to their young. Darwin and Wallace suggested that wild animal species also gradually evolved through natural selection, a similar system to the artificial selection that people were using to breed prize cattle, sheep and pedigree dogs.

The theory of natural selection is often described as the survival of the fittest. This is because animals must compete with each other for limited resources including food, water, shelter and mates. But they are not all equal or exactly similar, and some members of a population of animals will have characteristics which make them "fitter" – better suited to the environment at that time.

The fitter animals will therefore be more successful at finding food and avoiding predators. Consequently, they will probably produce more offspring, many of which will also have the same characteristics as their fit parents. Because of this, the next generation

Jumping animals

Most animals can leap into the air, but thanks to natural selection this simple ability has been harnessed by different animals in different ways. For example,

click beetles jump in somersaults to frighten off attackers, while blood-sucking fleas can leap enormous heights to move from host to host.

Above: The flying frog uses flaps of skin between its toes to glide. This allows these tree-living frogs to leap huge distances between branches.

Above: The bobcat is an agile mammal with powerful legs that allow it to leap over 3.6m (12ft) into the air to pounce on its prey, such as hares, porcupines, birds and even deer.

will contain more individuals with the "fit" trait. And after many generations, it is possible that the whole population will carry the fit trait, since those without it die out.

Variation and time

The environment is not fixed, and does not stay the same for long. Volcanoes, diseases and gradual climate changes, for example, alter the conditions which animals have to confront. Natural selection relies on the way in which different individual animals cope with these changes. Those individuals that were once fit may later die out, as others that have a different set of characteristics become more successful in the changed environment.

Darwin did not know it, but parents pass their features on to their young through their genes. During sexual reproduction, the genes of both parents are jumbled up to produce a new individual with a unique set of characteristics. Every so often the genes mutate into a new form, and these mutations are the source of all new variations.

As the process of natural selection continues for millions of years, so groups of animals can change radically, giving rise to a new species. Life is thought to have been evolving for 3.5 billion years. In that time natural selection has produced a staggering number of species, with everything from oak trees to otters and coral to cobras.

A species is a group of organisms that can produce offspring with each other. A new species occurs once animals have changed so much that they are unable to breed with their ancestors. And if the latter no longer exist, then they have become extinct.

New species may gradually arise out of a single group of animals. In fact the original species may be replaced by one or more new species. This can happen when two separate groups of one species are kept apart by an impassable geographical feature, such as an ocean or mountain range. Kept isolated from each other, both groups then evolve in different ways and end up becoming new species.

Mammalian evolution

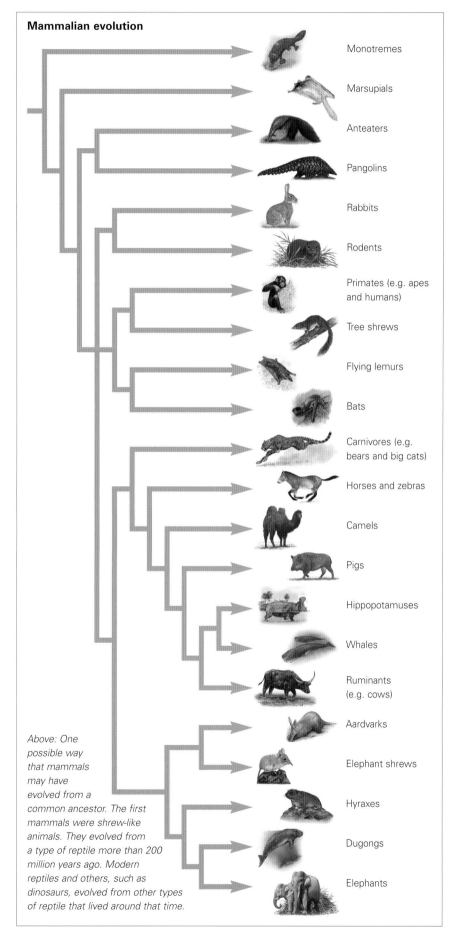

- Monotremes
- Marsupials
- Anteaters
- Pangolins
- Rabbits
- Rodents
- Primates (e.g. apes and humans)
- Tree shrews
- Flying lemurs
- Bats
- Carnivores (e.g. bears and big cats)
- Horses and zebras
- Camels
- Pigs
- Hippopotamuses
- Whales
- Ruminants (e.g. cows)
- Aardvarks
- Elephant shrews
- Hyraxes
- Dugongs
- Elephants

Above: One possible way that mammals may have evolved from a common ancestor. The first mammals were shrew-like animals. They evolved from a type of reptile more than 200 million years ago. Modern reptiles and others, such as dinosaurs, evolved from other types of reptile that lived around that time.

ANATOMY

Mammals, reptiles and amphibians (which are vertebrates, as are fish and birds), come in a mind-boggling array of shapes and sizes. However all of them, from whales to bats and frogs to snakes, share a basic body plan, both inside and out.

Vertebrates are animals with a spine, generally made of bone. Bone, the hard tissues of which contain chalky substances, is also the main component of the rest of the vertebrate skeleton. The bones of the skeleton link together to form a rigid frame to protect organs and give the body its shape, while also allowing it to move. Cartilage, a softer, more flexible but tough tissue is found, for example, at the ends of bones in mobile joints, in the ears and the nose (forming the sides and the partition between the two nostrils). Some fish, including sharks and rays, have skeletons that consist entirely of cartilage.

Nerves and muscles

Vertebrates also have a spinal cord, a thick bundle of nerves extending from the brain through the spine, and down into the tail. The nerves in the spinal cord are used to control walking and other reflex movements. They also coordinate blocks of muscle that work together for an animal to move properly. A vertebrate's skeleton is on the inside, in contrast to many invertebrates, which have an outer skeleton or exoskeleton.

The vertebrate skeleton provides a solid structure which the body's muscles pull against. Muscles are blocks of protein that can contract and relax when they get an electrical impulse from a nerve.

Invertebrates

The majority of animals are invertebrates. They are a much more varied group than the vertebrates and include creatures as varied as shrimps, slugs, butterflies and starfish. Although some squid are thought to reach the size of a small whale, and while octopuses are at least as intelligent as cats and dogs, most invertebrates are much smaller and simpler animals than the vertebrates.

Below: The most successful invertebrates are the insects, including ants. This soldier army ant is defending workers as they collect food.

Reptile bodies

Reptiles have internal skeletons made from bone and cartilage. Their skins are covered in scales, which are often toughened by a waxy protein called keratin. Turtles are quite different from other reptiles. They have simpler skulls and a shell that is joined to the animal's internal skeleton.

Below: Crocodiles have very strong bodies, designed for life in and around shallow water.

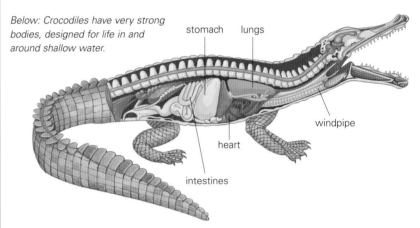

stomach
lungs
heart
windpipe
intestines

Below: Lizards have a similar body plan to crocodiles, although they are actually not very closely related.

Below: Snakes' internal organs are elongated so that they fit into their long, thin bodies. One of a pair of organs, such as the lungs, is often very small or missing.

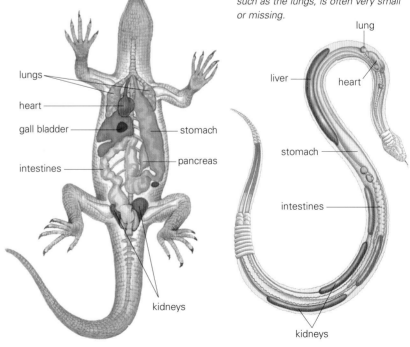

lungs
heart
gall bladder
intestines
stomach
pancreas
kidneys

lung
liver
heart
stomach
intestines
kidneys

When on the move, the vertebrate body works like a system of pulleys, pivots and levers. The muscles are the engines of the body, and are attached to bones – the levers – by strong cables called tendons. The joint between two bones forms a pivot, and the muscles work in pairs to move a bone. For example, when an arm is bent at the elbow to raise the forearm, the bicep muscle on the front of the upper arm has to contract. This pulls the forearm up, while the tricep muscle attached to the back of the upper arm remains relaxed. To straighten the arm again, the tricep contracts and the bicep relaxes. If both muscles contract at the same time, they pull against each other, and the arm remains locked in whatever position it is in.

Vital organs

Muscles are not only attached to the skeleton. The gut – including the stomach and intestines – is surrounded by muscles. These muscles contract in rhythmic waves to push food and waste products through the body. The heart is a muscular organ made of a very strong muscle which keeps on contracting and relaxing, pumping blood around the body. The heart and other vital organs are found in the thorax, that part of the body which lies between the forelimbs. In reptiles and mammals the thorax is kept well protected inside a rib cage, as are the lungs, liver and kidneys.

Vertebrates have a single liver consisting of a number of lobes. The liver has a varied role, making chemicals required by the body and storing food. Most vertebrates also have two kidneys. Their role is to clean the blood of any impurities and toxins, and to remove excess water. The main toxins that have to be removed are compounds containing nitrogen, the by-products of eating protein. Mammal and amphibian kidneys dissolve these toxins in water to make urine. However, since many reptiles live in very dry habitats, they cannot afford to use water to remove waste, and they instead get rid of it as a solid waste similar to bird excrement.

Mammalian bodies

Most mammals are four-limbed (exceptions being sea mammals such as whales). All have at least some hair on their bodies, and females produce milk. Mammals live in a wide range of habitats and their bodies are adapted in many ways to survive. Their internal organs vary depending on where they live and what they eat.

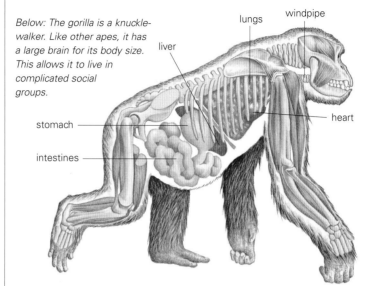

Below: The gorilla is a knuckle-walker. Like other apes, it has a large brain for its body size. This allows it to live in complicated social groups.

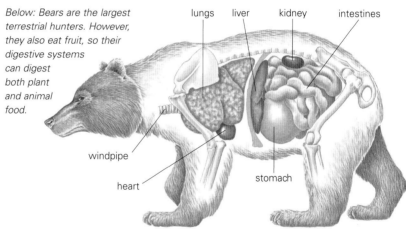

Below: Bears are the largest terrestrial hunters. However, they also eat fruit, so their digestive systems can digest both plant and animal food.

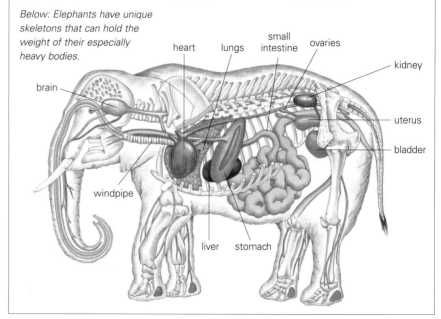

Below: Elephants have unique skeletons that can hold the weight of their especially heavy bodies.

SENSES

To stay alive, animals must find food and shelter, and defend themselves against predators. To achieve these things, they are equipped with an array of senses for monitoring their surroundings. Different species have senses adapted to nocturnal or diurnal (day-active) life.

An animal's senses are its early-warning system. They alert it to changes in its surroundings – changes which may signal an opportunity to feed or mate, or the need to escape imminent danger. The ability to act quickly and appropriately is made possible because the senses are linked to the brain by a network of nerves which send messages as electric pulses. When the brain receives the information from the senses it coordinates its response.

In many cases, generally in response to something touching the body, the signal from the sensor does not reach the brain before action is taken. Instead, it produces a reflex response which is hardwired into the nervous system. For example, when you touch a very hot object, your hand automatically recoils; you don't need to think about it.

All animals have to be sensitive to their environment to survive. Even the simplest animals, such as jellyfish and roundworms, react to changes in their surroundings. Simple animals, however, have only a limited ability to move or defend themselves, and therefore generally have limited senses. Larger animals, such as vertebrates, have a much more complex array of sense organs. Most vertebrates can hear, see, smell, taste and touch.

Vision

Invertebrates' eyes are generally designed to detect motion. Vertebrates' eyes, however, are better at forming clear images, often in color. Vertebrates' eyes are balls of clear jelly which have an inner lining of light-sensitive cells. This lining, called the retina, is made up of one or two types of cell. The rod cells – named after their shape – are very sensitive to all types of light, but are only capable of forming black and white images. Animals which are active at night generally have (and need) only rods in their eyes.

Color vision is important for just a few animals, such as monkeys, which need, for example, to see the brightest and therefore ripest fruits. Color images are made by the cone cells – so named because of their shape – in the retina. There are three types of cone, each of which is sensitive to a particular wavelength of light. Low wavelengths appear as reds, high wavelengths as blues, with greens being detected in between.

Above: Frogs have large eyes positioned on the upper side of the head so that the animals can lie mainly submerged in water with just their eyes poking out.

The light is focused on the retina by a lens to produce a clear image. Muscles change the shape of the lens so that it can focus the light arriving from different distances. While invertebrates may have several eyes, all vertebrates have just two, and they are always positioned on the head. Animals such as rabbits, which are constantly looking out for danger, have eyes on the side of the head to give a wide field of vision. But while they can see in almost all directions, rabbits have difficulty judging distances and speeds. Animals that have eyes pointing forward are better at doing this because each eye's field of vision overlaps with the other. This binocular vision helps hunting animals and others, such as tree-living primates, to judge distances more accurately.

Eyes can also detect radiation in a small band of wavelengths, and some animals detect radiation that is invisible to our eyes. Flying insects and birds can see ultraviolet light, which extends the range of their color vision. At the other end of the spectrum many snakes can detect radiation with a lower wavelength. They sense infrared, or heat, through pits on the face which enables them to track their warm-blooded prey in pitch darkness.

Below: The raccoon is a nocturnal animal with excellent night vision, and eyes that glow bright yellow in the dark. These animals have a distinctive black mask across their eyes.

Below: Like other hunters, a seal has eyes positioned on the front of its head. Forward-looking eyes are useful for judging distances, making it easier to chase down prey.

Hearing

An animal's brain interprets waves of pressure travelling through the air, and detected by the ears, as sound. Many animals do not hear these waves with ears but detect them in other ways instead. For example, although snakes can hear, they are much more sensitive to vibrations through the lower jaw, travelling through the ground. Long facial whiskers sported by many mammals, from cats to dugongs, are very sensitive touch receptors. They can be so sensitive that they will even respond to currents in the air.

In many ways, hearing is a sensitive extension of the sense of touch. The ears of amphibians, reptiles and mammals have an eardrum which is sensitive to tiny changes in pressure. An eardrum is a thin membrane of skin which vibrates as the air waves hit it. A tiny bone (or in the case of mammals, three bones) attached to the drum transmit the vibrations to a shell-shaped structure called a cochlea. The cochlea is filled with a liquid which picks up the vibrations. As the liquid moves inside the cochlea, tiny hair-like structures lining it wave back and forth. Nerves stimulated by this wave motion send the information to the brain, which interprets it as sound.

A mammal's ear is divided into three sections. The cochlea forms the inner ear and the middle ear consists of the bones between the cochlea and eardrum. The outer ear is the tube joining the outside world and the

Above: Snakes have forked tongues that they use to taste the air. The tips of the fork are slotted into an organ in the roof of the mouth. This organ is linked to the nose, and chemicals picked up by the tongue are identified with great sensitivity.

auricle – the fleshy structure on the side of the head that collects the sound waves – to the middle ear. Amphibians and reptiles do not possess auricles. Instead their eardrums are either on the side of the head – easily visible on many frogs and lizards – or under the skin, as in snakes.

Smell and taste

Smell and taste are so closely related as to form a single sense. Snakes and lizards, for example, taste the air with their forked tongues. However, it is perhaps the most complex sense. Noses, tongues and other smelling

organs are lined with sensitive cells which can analyze a huge range of chemicals that float in the air or exist in food. Animals such as dogs, which rely on their sense of smell, have long noses packed with scent-sensitive cells. Monkeys, of the other hand, are less reliant on a sense of smell, and consequently have short noses capable only of detecting stronger smells.

Below: Hares have very large outer ears which they use like satellite dishes to pick up sound waves. They can rotate each ear separately to detect sound from all directions.

Below: Lizards do not have outer ears at all. Their hearing organs are contained inside the head and joined to the outside world through an eardrum membrane.

Below: Wolves have an excellent sense of smell and taste. They communicate with pack members and rival packs by smell, as part of a complex set of social attitudes.

SURVIVAL

In order to stay alive, animals must not only find enough food, but also avoid becoming a predator's meal. To achieve this, animals have evolved many strategies to feed on a wide range of foods, and an array of weapons and defensive tactics to keep safe.

An animal must keep feeding in order to replace the energy used in staying alive. Substances in the food, such as sugars, are burned by the body, and the subsequent release of energy is used to heat the body and power its movements. Food is also essential for growth. Although most growth takes place during the early period of an animal's life, it never really stops because injuries need to heal and worn-out tissues need replacing. Some animals continue growing throughout life. Proteins in the food are the main building blocks of living bodies.

Plant food

Some animals will eat just about anything, while others are much more fussy. As a group, vertebrates get their energy from a wide range of sources – everything from shellfish and wood to honey and blood. Animals are often classified according to how they feed, forming several large groups filled with many otherwise unrelated animals.

Animals that eat plants are generally grouped together as herbivores. But this term is not very descriptive because there is such a wide range of plant foods. Animals that eat grass are known as grazers. However, this term can also apply to any animal which eats any plant that covers the ground

Above: Bison are grazers. They eat grass and plants that grow close to the ground. Because their food is all around them, grazers spend a long time out in the open. They feed together in large herds since there is safety in numbers.

in large amounts, such as seaweed or sedge. Typical grazers include bison and wildebeest but some, such as the marine iguana or gelada baboon, are not so typical. Animals such as giraffes or antelopes, which pick off the tastiest leaves, buds and fruit from bushes and trees, are called browsers. Other browsing animals include many monkeys, but some monkeys eat only leaves (the folivores) or fruit (the frugivores).

Many monkeys have a much broader diet, eating everything from insects to the sap which seeps out from the bark of tropical trees. Animals that eat both plant and animal foods are called omnivores. Bears are omnivorous, as are humans, but the most catholic of tastes belong to scavenging animals, such as rats and other rodents, which eat anything they can get their teeth into. Omnivores in general, and scavengers in particular, are very curious animals. They will investigate anything that looks or smells like food, and if it also tastes like food, then it probably is.

A taste for flesh

The term carnivore is often applied to any animal that eats flesh, but it is more correctly used to refer to an order of mammals which includes cats, dogs, bears and many smaller animals, such as weasels and mongooses. These animals are the kings of killing, armed with razor-sharp claws and powerful jaws crammed full of chisel-like teeth. They use their strength and speed to overpower their prey, either by running them down or taking them by surprise with an ambush.

Below: Amazon squirrel monkeys feed on small insects, soft fruits and the nectar from flowers. Their highly dexterous fingers are ideally suited to searching for this kind of food.

Below: Grizzly bears are omnivores, eating both vegetation and animals. When salmon are abundant, they congregate in groups to share this protein-rich resource.

Above: Wolverines are primarily meat-eaters, capable of bringing down prey that are five times bigger than themselves. Their diet includes beavers, squirrels, marmots, rabbits, deer, moose, wild sheep and carrion.

However, land-dwelling carnivores are not the only expert killers. The largest meat-eater is the orca, or killer whale, which is at least three times the size of the brown bear, the largest killer on land.

While snakes are much smaller in comparison, they are just as deadly, if not more so. They kill in one of two ways, either suffocating their prey by wrapping their coils tightly around them, or by injecting them with a poison through their fangs.

Arms race

Ironically, the same weapons used by predators are often used by their prey to defend themselves. For example, several species of frog, toad and salamander secrete poisons on to their skin. In some cases, such as the poison-dart frog, this poison is enough to kill any predator that tries to eat it, thus making sure that the killer won't repeat its performance. More often, though, a predator finds that its meal tastes horrible and remembers not to eat another one again. To remind the predators to keep away, many poisonous amphibians are bright green,

red, yellow, orange or blue, which ensures that they are easily recognized.

Many predators rely on stealth to catch their prey, and staying hidden is part of the plan. A camouflaged coat, such as a tiger's stripes, helps animals blend into their surroundings. Many species also use this technique to ensure that they do not get eaten. Most freeze when danger approaches, and then scurry to safety as quickly as possible. Chameleons have taken camouflage to an even more sophisticated level as they can change the hue of their scaly skins, which helps them to blend in with their surrounding environment.

Plant-eating animals that live in the open cannot hide from predators that are armed with sharp teeth and claws. And the plant-eaters cannot rely on similar weapons to defend themselves. They are outgunned

Right: Chameleons have skin cells that can be opened and closed to make their skin tint change.

Filter-feeders

Some animals filter their food from water. The giant baleen whales do this, sieving tiny shrimp-like animals called krill out of great gulps of sea water. Some tadpoles and larval salamanders filter-feed as well, extracting tiny plant-like animals which float in fresh water. However, after becoming adults, all amphibians become hunters and eat other animals. All snakes and most lizards are meat-eaters, or carnivores, as well.

Below: The largest animals of all, baleen whales, are filter-feeders. They do not have teeth. Instead, their gums are lined with a thick curtain of baleen that filters out tiny shrimp-like krill from sea water.

because they do not possess sharp, pointed teeth but flattened ones to grind up their plant food. The best chance they have of avoiding danger is to run away. Animals such as antelopes or deer consequently have long, hoofed feet that lengthen their legs considerably; they are, in fact, standing on their toenails. These long legs allow them to run faster and leap high into the air to escape an attacker's jaws.

Animals that do not flee must stand and fight. Most large herbivores are armed with horns or antlers. Although used chiefly for display, the horns are the last line of defence when cornered.

REPRODUCTION

All animals share the urge to produce offspring which will survive after the parents die. The process of heredity is determined by genes, through which characteristics are passed from parents to offspring. Reproduction presents several problems, and animals have adopted different strategies for tackling them.

Animals have two main goals: to find food and a mate. To achieve these goals, they must survive everything that the environment throws at them, from extremes of the weather, such as floods and droughts, to hungry predators. They have to find sufficient supplies of food, and on top of that locate a mate before their competitors. If they find sufficient food but fail to produce any offspring, their struggle for survival will have been wasted.

One parent or two?

There are two ways in which an animal can reproduce, asexually or sexually. Animals that are produced by asexual reproduction, or parthenogenesis, have only one parent, a mother. The offspring are identical to their mother and to each other. Sexual reproduction involves two parents of the opposite sex. The offspring are hybrids of the two parents, with a mixture of their parents' characteristics.

The offspring inherit their parents' traits through their genes. Genes can be defined in various ways. One simple definition is that they are the unit of inheritance – a single inherited

Below: Crocodiles bury their eggs in a nest. The temperature of the nest determines the sex of the young reptiles. Hot nests produce more males than cool ones. Crocodile mothers are very gentle when it comes to raising young.

Above: Many male frogs croak by pumping air into an expandable throat sac. The croak is intended to attract females. The deeper the croak, the more attractive it is. However, some males lurk silently and mate with females as they approach the croaking males.

characteristic which cannot be subdivided any further. Genes are also segments of DNA (deoxyribonucleic acid), a complex chemical that forms long chains. It is found at the heart of every living cell. Each link in the DNA chain forms part of a code that controls how an animal's body develops and survives. And every cell in the body contains a full set of DNA which could be used to build a whole new body.

Animals produced through sexual reproduction receive half their DNA, or half their genes, from each parent. The male parent provides half the supply of genes, contained in a sperm. Each sperm's only role is to find its

Above: In deer and many other grazing animals, the males fight each other for the right to mate with the females in the herd. The deer with the largest antlers often wins without fighting, and real fights only break out if two males appear equally well-endowed.

way to, and fertilize, an egg, its female equivalent. Besides containing the other half of the DNA, the egg also holds a supply of food for the offspring as it develops into a new individual. Animals created through parthenogenesis get all their genes from their mother, and all of them are therefore the same sex – female.

Pros and cons

All mammals reproduce sexually, as do most reptiles and amphibians. However, there are a substantial number of reptiles and amphibians, especially lizards, which reproduce by parthenogenesis. There are benefits and disadvantages to both types of reproduction. Parthenogenesis is quick and convenient. The mother does not need to find a mate, and can devote all of her energy to producing huge numbers of young. This strategy is ideal for populating as yet unexploited territory. However, being identical, these animals are very vulnerable to attack. If, for example, one is killed by a disease or outwitted by a predator, it is very likely that they will all suffer the same fate. Consequently, whole communities of animals produced through parthenogenesis can be wiped out.

Sexual animals, on the other hand, are much more varied. Each one is unique, formed by a mixture of genes from both parents. This variation means that a group of animals produced by sexual reproduction is more likely to triumph over adversity than a group of asexual ones. However, sexual reproduction takes up a great deal of time and effort.

Attracting mates

Since females produce only a limited number of eggs, they are keen to make sure that they are fertilized by a male with good genes. If a male is fit and healthy, this is a sign that he has good genes. Good genes will ensure that the offspring will be able to compete with other animals for food and mates of their own. Because the females have the final say in agreeing to mate, the

Above: Prairie dogs usually have one litter of four to six young per year. The young are born hairless and helpless, with their eyes closed. They remain underground for the first six weeks of life, then emerge from their dens.

Below: Male moose compete for a harem of females by sparring with their antlers. This fighting rarely results in injuries, but sometimes antlers may break or become entangled.

Above: Grizzly bear cubs are blind and helpless at birth. Their mother's milk is rich in fat and calories, so the cubs develop quickly. Mothers are fiercely protective of their young, defending them against wolves and cougars. Cubs stay with their mothers for up to three years.

males have to put a lot of effort into getting noticed. Many are brightly colored, make loud noises, and they are often larger than the females. In many species the males even compete with each other for the right to display to the females. Winning that right is a good sign that they have the best genes.

Parental care

The amount of care that the offspring receive from their parents varies considerably. There is a necessary trade-off between the amount of useful care parents can give to each offspring, the number of offspring they can produce and how regularly they can breed. Mammals invest heavily in parental care, suckling their young after giving birth, while most young amphibians or reptiles never meet their parents at all.

By suckling, mammals ensure that their young grow to a size where they can look after themselves. Generally, the young stay with the mother until it is time for her to give birth to the next litter – at least one or two months. However, in many species, including humans, the young stay with their parents for many years.

Other types of animals pursue the opposite strategy, producing large numbers of young that are left to fend for themselves. The vast majority in each batch of eggs – consisting of hundreds or even thousands – die before reaching adulthood, and many never even hatch. The survival rates, for example of frogs, are very low.

Animals that live in complicated societies, such as elephants, apes and humans, tend to produce a single offspring every few years. The parents direct their energies into protecting and rearing the young, giving them a good chance of survival. Animals which live for a only a short time, such as mice, rabbits, and reptiles and amphibians in general, need to reproduce quickly to make the most of their short lives. They produce high numbers of young, and do not waste time on anything more than the bare minimum of parental care. If successful, these animals can reproduce at an alarming pace.

AMPHIBIANS

Amphibians are the link between fish and land animals. One in eight of all vertebrate animals are amphibians. This group includes frogs, toads and newts as well as rarer types, such as giant sirens, hellbenders and worm-like caecilians. Amphibians are equally at home in water and on land.

Amphibians live on every continent except for Antarctica. None can survive in salt water, although a few species live close to the sea in the brackish water at river mouths. Being cold-blooded – their body temperature is always about the same as the temperature of their surroundings – most amphibians are found in the warmer regions of the world.

Unlike other land vertebrates, amphibians spend the early part of their lives in a different form from that of the adults. As they grow, the young gradually metamorphose into the adult body. Having a larval form means that the adults and their offspring live and feed in different places. In general the larvae are aquatic, while the adults spend most of their time on land.

The adults are hunters, feeding on other animals, while the young are generally plant eaters, filtering tiny plants from the water or grazing on aquatic plants which line the bottom of ponds and rivers.

Below: Amphibians must lay their eggs near a source of water. In most cases, such as this frog spawn, the eggs are laid straight into a pond or swamp. The tadpoles develop inside the jelly-like egg and then hatch out after the food supply in the egg's yolk runs out.

Above: Amphibians begin life looking very different from the adult form. Most of the time these larval forms, such as this frog tadpole, live in water as they slowly develop into the adult form, growing legs and lungs so that they can survive on land.

Life changing

Most amphibians hatch from eggs laid in water or, in a few cases, in moist soil or nests made of hardened mucus. Once hatched, the young amphibians, or larvae, live as completely aquatic animals. Those born on land wriggle to the nearest pool of water or drop from their nest into a river.

The larvae of frogs and toads are called tadpoles. Like the young of salamanders – a group that includes all other amphibians except caecilians – tadpoles do not have any legs at first. They swim using a long tail that has a fish-like fin extending along its length. As they grow, the larvae sprout legs. In general the back legs begin to grow first, followed by the front pair. Adult frogs do not have tails, and after the first few months a tadpole's tail begins to be reabsorbed into the body – it does not just fall away.

All adult salamanders keep their tails, and those species that spend their entire lives in water often retain the fin along the tail, along with other parts, such as external gills, a feature that is more commonly seen in the larval stage.

Body form

Amphibian larvae hatch with external gills but, as they grow, many (including all frogs and the many salamanders which live on land) develop internal gills. In most land-living species these internal gills are eventually replaced by lungs. Amphibians are also able to absorb oxygen directly through the skin, especially through the thin and moist tissues inside the mouth. A large number of land-living salamanders get all their oxygen in this way because they do not have lungs.

All adult frogs and toads return to the water to breed and lay their eggs, which are often deposited in a jelly-like mass called frog spawn. Several types of salamander do not lay eggs, and instead the females keep the fertilized eggs inside their bodies. The larvae develop inside the eggs, fed by a supply of rich yoke, and do not hatch until they have reached adult form.

Above: After the first few weeks, a tadpole acquires tiny back legs. As the legs grow, the long tail is gradually reabsorbed into the body. The front legs appear after the back ones have formed.

Adult form

Most adult amphibians have four limbs, with four digits on the front pair and five on the rear. Unlike other land-living animals, such as reptiles or mammals, their skin is naked and soft. Frogs' skin is smooth and moist, while toads generally have a warty appearance.

The skins of many salamanders are brightly colored, with patterns that often change throughout the year. Color change prior to the mating season signals the salamander's readiness to mate. Many frogs also have bright skin colors. Although their skin shades can change considerably in different light levels, these are generally not mating signals to fellow frogs. Instead they are warnings to predators that the frog's skin is laced with deadly poison. While toads tend to be drab in color, many also secrete toxic chemicals to keep predators away. These substances are often stored in swollen warts which burst when the toad is attacked.

Below: Adult frogs may live in water or on land. Aquatic ones have webbed feet, while those on land have powerful legs for jumping and climbing. All frogs must return to a source of water to mate.

Forever young

Salamanders which have changed into adults, but which have not yet reached adult size, are called efts. The time it takes for an amphibian to grow from a newly hatched larva to an adult varies considerably, and the chief factor is the temperature of the water in which it is developing. Most frogs and toads develop in shallow waters, warmed by the summer sun, and they generally reach adulthood within three to four months. However, salamanders, especially the largest ones, can take much longer, and at the northern and southern limits of their geographical spread some salamanders stay as larvae for many years. It appears that the trigger for the change into adult form is linked to the temperature, and in cold climates this change happens only every few years. In fact it may not happen during a salamander's lifetime, and consequently several species have evolved the ability to develop sexual organs even when they still look like larvae.

Below: Marbled salamanders are unusual in that the females lay their eggs on dry land and coil themselves around them to keep them as moist as possible. They stay like this until the seasonal rains fall. The water stimulates the eggs to hatch.

REPTILES

Reptiles include lizards, snakes, alligators, crocodiles, turtles and tortoises, as well as now-extinct creatures such as dinosaurs and the ancestors of birds. Crocodiles have roamed the Earth for 200 million years and are still highly successful hunters.

Reptiles are a large and diverse group of animals containing about 6,500 species. Many of these animals look very different from each other and live in a large number of habitats, from the deep ocean to the scorching desert. Despite their great diversity, all reptiles share certain characteristics.

Most reptiles lay eggs, but these are different from those of an amphibian because they have a hard, thin shell rather than a soft, jelly-like one. This protects the developing young inside and, more importantly, stops them from drying out. Shelled eggs were an evolutionary breakthrough because they meant that adult reptiles did not have to return to the water to breed. Their waterproof eggs could be laid anywhere, even in the driest places. Reptiles were also the first group of land-living animals to develop into an adult form inside the egg. They did not emerge as under-developed larvae like the young of most amphibians.

Below: Alligators and other crocodilians are an ancient group of reptiles that have no close living relatives. They are archosaurs, a group of reptiles that included the dinosaurs. Other living reptiles belong to a different group.

Released from their ties to water, the reptiles developed unique ways of retaining moisture. Their skins are covered by hardened plates or scales to stop water being lost. The scales are also coated with the protein keratin, the same substance used to make fingernails and hair.

All reptiles breathe using lungs; if they were to absorb air through the skin it would involve too much water loss. Like amphibians, reptiles are cold-blooded and cannot heat their bodies from within as mammals can. Consequently, reptiles are commonly found in warm climates.

Ancient killers

Being such a diverse group, reptiles share few defining characteristics besides their shelled eggs, scaly skin and lungs. They broadly divide into four orders. The first contains the crocodiles, and includes alligators and caimans; these are contemporaries of the dinosaurs, both groups being related to a common ancestor.

In fact today's crocodiles have changed little since the age when dinosaurs ruled the world over 200 million years

Above: Turtles and their relatives, such as these giant tortoises, are unusual reptiles. Not only do they have bony shells fused around their bodies, but they also have skulls that are quite different from other reptiles. Turtles are also unusual because many of them live in the ocean, while most reptiles live on land.

ago. Unlike the dinosaurs, which disappeared 65 million years ago, the crocodiles are still going strong. Technically speaking, the dinosaurs never actually died out; their direct descendents, the birds, are still thriving. Although birds are now grouped separately from reptiles, scientists know that they all evolved from ancestors which lived about 400 million years ago. Mammals, on the other hand, broke away from this group about 300 million years ago.

Above: Most reptiles, including this tree boa, lay eggs. The young hatch looking like small versions of the adults. However, several snakes and lizards give birth to live young, which emerge from their mother fully formed.

Distant relatives

The second reptile order includes turtles, terrapins and tortoises. These are only distantly related to other reptiles, and it shows. Turtles are also the oldest group of reptiles, evolving their bony shells and clumsy bodies before crocodiles, dinosaurs or any other living reptile group existed. Although turtles evolved on land, many have since returned to water. However, they still breathe air, and all must return to land to lay their eggs.

The third group of reptiles is the largest. Collectively called the squamates, this group includes snakes and lizards.

Snakes, with their legless bodies and formidable reputations, are perhaps the most familiar reptiles. They evolved from animals that did have legs, and many retain tiny vestiges of legs. The squamates include other legless members such as the amphisbaenians (or worm lizards) and slow worms. Both of these groups are more closely related to lizards than snakes, despite looking more like the latter. Lizards are not a simple group of reptiles, and many biologists refer to them as several different groups, including the skinks, monitors, geckos and iguanas.

Below: Lizards, such as this iguana, are the largest group of reptiles. Most are hunters that live in hot parts of the world, and they are especially successful in dry areas where other types of animal are not so common.

The squamates are so diverse in their lifestyles and body forms that it is hard to find factors which they have in common. One feature not found in other reptile orders is the Jacobson's organ. It is positioned in the roof of the mouth and is closely associated with the nose. All snakes and most lizards use this organ to "taste" the air and detect prey. The long forked tongue of most of these animals flicks out into the air, picking up tiny particles on its moist surface. Once back inside the mouth, each fork slots into the Jacobson's organ which then analyzes the substances.

The fourth and final order of reptiles is very small: the tuataras. These include just a few species, all of which are very rare indeed, clinging to life on islands off the mainland of New Zealand. To most people a tuatara looks like a large iguana. However, scientists believe that it is only a distant relative of lizards and other squamates because it has an odd-shaped skull and no eardrums.

Tuatara

Despite their differences from lizards and other squamates, tuataras do share one feature with lizards: the so-called third eye. This light-sensitive gland inside the head can detect light penetrating the thin skull. Both types of reptile use the third eye to regulate their exposure to the sun and so regulate their body temperature throughout the year.

Below: The tuatara is a living fossil, living on just a few islands around New Zealand. It looks like a lizard, but its skull and skeleton show that it is the last member of another ancient group of reptiles.

PLACENTAL MAMMALS

Mammals are the most familiar of all vertebrates. This is because not only are human beings mammals, but also most domestic animals and pets belong in this category. Placental mammals are also more widespread than other types of animal, being found in all parts of the world.

Mammals are grouped together because they share a number of characteristics. However, these common features do not come close to describing the huge diversity within the mammal class. For example, the largest animal that has ever existed on Earth – the blue whale – is a mammal, and so this monster of the deep shares several crucial traits with even the smallest mammals, such as the tiniest of shrews. Other mammals include elephants and moles, monkeys and hippopotamuses, and bats and camels. To add to this great diversity, mammals live in more places on Earth than any other group of animals, from the frozen ice fields of the Arctic to the humid treetops of the Amazon rainforest, and even under the sandy soil of African deserts.

Above: With their thick brown fur and powerful clawed paws, bears are very well adapted to life at the top of the food chain. All bears have a similar basic body plan, which is what makes them such successful terrestrial hunters.

Mammal bodies

The most obvious mammalian feature is hair. All mammals have hair made of keratin protein and, in most cases, it forms a thick coat of fur, though many mammals are relatively naked, not least humans. Unlike reptiles and amphibians, all mammals are warm-blooded, which means that they can keep their body temperature at a constant level. While this requires quite a lot of energy, it means that mammals are not totally dependent on the temperature of their surroundings. In places where other vertebrates would be frozen solid, mammals can survive by seeking out food and keeping themselves warm. Many mammals, including humans, can also cool their bodies using sweat. The water secreted on the skin cools the body efficiently, but it does mean that these animals need to drink more replacement water than do other groups.

Incidentally, the name mammal comes from the mammary glands. These glands are the means by which all female mammals provide milk (or liquid food) to their developing young. The young suck the milk through teats or nipples for the first few weeks or months of life.

Reproduction

Mammals reproduce in a number of ways. Monotremes, such as the duck-billed platypus, lay eggs, but all other mammals give birth to their young. Marsupials, a relatively small group of animals which includes kangaroos, give birth to very undeveloped young which then continue to grow inside a fold or pouch on the mother's skin.

Below: Although they are often mistaken for fish, dolphins are mammals: they breathe air and suckle their young. However, life under water requires flippers and fins, not legs.

The majority of mammals, called the placental mammals or eutherians, do not give birth to their young until they are fully formed and resemble the adults. The developing young, or foetuses, grow inside a womb or uterus where they are fed by the mother through a placenta. This large organ allows the young to stay inside the mother for a lot longer than in most other animals. It forms the interface between the mother's blood supply and that of the developing foetus, where oxygen and food pass from the parent to her offspring. The placenta is attached to the foetus by means of an umbilical cord which withers and drops off soon after the birth.

Widespread range

Mammals are found in a wider variety of habitats than any other group of animals. While mammals all breathe air with their

Right: One factor that makes mammals unique is that the females have mammary glands. These glands produce milk for the young animals to drink, as this fallow deer fawn is doing. The milk is a mixture of fat, protein and sugars.

Above: Female golden lion tamarins usually give birth to twins, but all the adults in the group cooperate to carry and feed the young, with the adult males doing the largest share.

lungs, this has not prevented many from making their homes in water. In many ways the streamlined bodies of whales and dolphins, for example, resemble those of sharks and other large fish. However, they are very much mammals, breathing air through a large nostril or blowhole in the top of the head, but their body hair has been reduced to just a few thick bristles.

Above: Plenty of mammals can glide, but only bats join birds and insects in true flight. A bat wing is made from skin that is stretched between long finger bones.

At the other end of the spectrum, some mammals even fly. Bats darting through the gloom of a summer evening may appear to be small birds, but they too are mammals with furry bodies and wings made from stretched skin instead of feathers. Although most other mammals have a more conventional body plan, with four legs and a tail, they too have evolved to survive in a startling range of habitats. They have achieved this not just by adapting their bodies but by changing how they behave. In general, mammals have larger brains than reptiles and amphibians, and this allows them to understand their environment more fully. Many mammals, such as monkeys and dogs, survive by living in complex social groups in which individuals cooperate with each other when hunting food, protecting the group from danger and even finding mates.

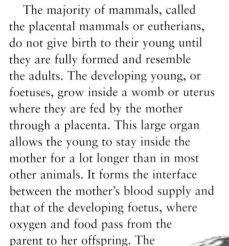

ECOLOGY

Ecology is the study of how groups of organisms interact with members of their own species, other organisms and the environment. All types of animals live in a community of interdependent organisms called an ecosystem, in which they have their own particular role.

The natural world is filled with a wealth of opportunities for animals to feed and breed. Every animal species has evolved to take advantage of a certain set of these opportunities, called a niche. A niche is not just a physical place but also a lifestyle exploited by that single species. For example, even though they live in the same rainforest habitat, sloths and tapirs occupy very different niches.

To understand how different organisms interrelate, ecologists combine all the niches in an area into a community, called an ecosystem. Ecosystems do not really exist because it is impossible to know where one ends and another begins, but the system is a useful tool when learning more about the natural world.

Food chains

One way of understanding how an ecosystem works is to follow the food chains within it. A food chain is made up of a series of organisms that prey on each other. Each habitat is filled with them, and since they often merge into and converge from each other, they are often combined into food webs.

Below: Nature creates some incredible alliances. The American badger, for example, goes on hunting trips with a coyote. The coyote sniffs out the prey, and the badger digs it out of its burrow for both of them to eat.

Ecologists use food chains to see how energy and nutrients flow through natural communities. Food chains always begin with plants. Plants are the only organisms on Earth that do not need to feed, deriving their energy from sunlight, whereas all other organisms, including animals, get theirs from food. At the next level up the food chain come the plant-eaters. They eat the plants, and extract the sugar and other useful substances made by them. And, like the plants, they use these substances to power their bodies and stay alive. The predators occupy the next level up, and they eat the bodies of the plant-eating animals.

At each stage of the food chain, energy is lost, mainly as heat given out by the animals' bodies. Because of this, less energy is available at each level up the food chain. This means that in a healthy ecosystem there are always fewer predators than prey, and always more plants than plant-eaters.

Nutrient cycles

A very simple food chain would be as follows: grass, wildebeest and lion. However, the reality of most ecosystems is much more complex, with many more layers, including certain animals that eat both plants and animals. Every food chain ends with a top predator, in our example, the lion. Nothing preys on the lion, at least when it is alive, but once it dies the food chain continues as insects, fungi and other decomposers feed on the carcass. Eventually nothing is left of the lion's body. All the energy stored in it is removed by the decomposers, and the chemicals which made up its body have returned to the environment as carbon dioxide gas, water and minerals in the soil. And these are the very same substances needed by a growing plant. The cycle is complete.

Above: Nothing is wasted in nature. The dung beetle uses the droppings of larger grazing animals as a supply of food for its developing young. Since the beetles clear away all the dung, the soil is not damaged by it, the grass continues to grow, and the grazers have plenty of food.

Living together

As food chains show, the lives of different animals in an ecosystem are closely related. If all the plants died for some reason, it would not just be the plant-eaters that would go hungry. As all of them begin to die, the predators would starve too. Only the decomposers might appear to benefit. Put another way, the other species living alongside an animal are just as integral to that animal's environment as the weather and landscape. This is yet another way of saying that animal species have not evolved isolated from each another.

The result is that as predators have evolved new ways of catching their prey, the prey has had to evolve new ways of escaping. On many occasions this process of co-evolution has created symbiotic relationships between two different species. For example, honeyguide birds lead badgers to bees' nests.

Some niches are very simple, and the animals that occupy them live simple, solitary lives. Others, especially those occupied by mammals, are much more complex and require members of a species to live closely together. These aggregations of animals may be simple herds or more structured social groups.

Food chain

Food chains show how the energy needed for life passes through an ecosystem. The energy originates in the Sun. This makes plants grow, which are then eaten by animals. The plant-eating animals then become meals themselves.

Below: This food chain shows what animals eat in a temperate region, such as the majority of the USA. Herbivores eat only plants, while carnivores eat mainly other animals. Animals that eat both plants and animals are omnivores – for example, humans.

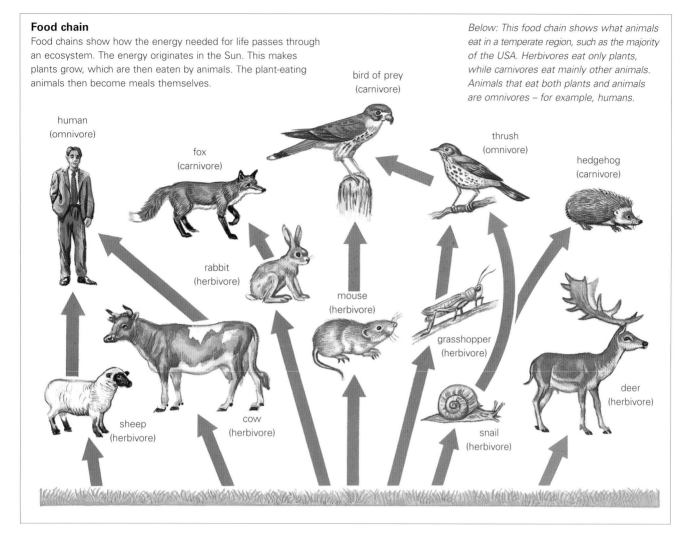

- human (omnivore)
- fox (carnivore)
- bird of prey (carnivore)
- thrush (omnivore)
- hedgehog (carnivore)
- rabbit (herbivore)
- mouse (herbivore)
- grasshopper (herbivore)
- deer (herbivore)
- sheep (herbivore)
- cow (herbivore)
- snail (herbivore)

Group living

A herd, flock or shoal is a group of animals which gathers together for safety. Each member operates as an individual, but is physically safest in the centre of the group, the danger of attack being greatest on the edge. Herd members do not actively communicate dangers to each other. When one is startled by something and bolts, the rest will probably follow.

Members of a social group, on the other hand, work together to find food, raise their young and defend themselves. Many mammals, for example apes, monkeys, dogs, dolphins and elephants, form social groups, and these groups exist in many forms. At one end of the spectrum are highly ordered societies, such as lion prides and baboon troops, which are often controlled by one dominant male, the other members often having their own ranking in a strict hierarchical structure. At the other end of the spectrum are leaderless gangs of animals, such as squirrel monkeys, which merge and split with no real guiding purpose.

There are many advantages of living in social groups, with members finding more food and being warned of danger, for example. However, in many societies only a handful of high-ranking members are allowed to breed. In these cases, the groups are held together by a complex fusion of family ties in which brothers and sisters help to raise nephews and nieces. Politics also plays its cohesive part, with members forming and breaking alliances in order to rise to the top.

Below: Bison have poor eyesight but good hearing and sense of smell. Living in vast herds gives them a better chance of detecting wolves. If attacked, bulls often form a defensive circle around their females and young.

MIGRATION AND HIBERNATION

Migration and hibernation are two ways in which animals cope with the changing seasons and fluctuations in the supply of food. By hibernating, they sleep through periods of bad weather when food is hard to find, and by migrating, they reach places where food is more readily available.

Everywhere on Earth, the climate changes throughout the year with the cycle of seasons. In some places these changes are hardly noticeable from month to month, while in others each new season brings extremes of weather from blistering hot summers to freezing winters, or torrential rains followed by drought.

Change of lifestyle

In temperate regions, such as Europe, the year is generally divided into four seasons. Other regions experience a different annual cycle of changes. For example, tropical regions do not really have fluctuating temperatures, but many areas do experience periods of relative dry and at least one period of heavier rains each year. By contrast, in the far north, the change between the short summer and long winter is so quick that, in effect, there are only two seasons.

Hibernating heart rate

The hibernating animal's heart rate slows to just a few beats per minute. It breathes more slowly and its body temperature drops to just a few degrees above the surrounding air temperature.

Below: The bodies of true hibernators, such as the dormouse, shut down almost completely during hibernation. Other hibernators, such as bears, may be out of sight for most of the winter, but they do not become completely dormant and their temperature does not fall drastically.

Animals must, of course, react to these changes if they are to survive the harshest weather and make the most of clement conditions. Monkeys, for example, build up a mental map of their patch of forest so that they know where the fresh leaves will be after the rains, and where to find the hardier forest fruits during a drought. Wolves living in chilly northern forests hunt together in packs during the cold winter, working co-operatively to kill animals which are much larger than they are. However, when the summer arrives they tend to forage alone, picking off the many smaller animals, such as rodents and rabbits, which appear when the snow melts.

Hibernation

The reason the wolves find these smaller animals in the summer is that they suddenly emerge having passed the winter in cosy burrows or nests. This is commonly called hibernating, but there is a distinction between true hibernation and animals simply being inactive over winter.

Animals such as bears and tree squirrels are not true hibernators. Although they generally sleep for long periods during the coldest parts of winter, hunkered down in a den or

Above: Reptiles that live in cooler parts of the world – rattlesnakes, for example – spend a long time lying dormant. They do not hibernate like mammals, but because they are cold-blooded and do not need lots of energy to function, they can go for long periods without food.

drey, they do not enter the deep, unconscious state of hibernation. Unable to feed while asleep, these animals rely on their bodily reserves of fat to stay alive. However, they often wake up during breaks in the harshest weather and venture outside to urinate or snatch a meal. Because tree squirrels have less fat reserves than bears, they frequently visit caches (stores) of food which they filled in the autumn.

On the other hand, the true hibernators, such as dormice or woodchucks, do not stir at all over winter. They rely completely on their reserves of fat to stay alive, and to save energy their metabolism slows to a fraction of its normal pace.

Only warm-blooded animals hibernate because they are the main types of animals which can survive in places where hibernation is necessary. However, rattlesnakes often pass the winter in rocky crevices and burrows. Reptiles and amphibians which live in very hot and dry places have their own form of hibernation, called aestivation.

Above: Whales have the longest migrations of all mammals. They move from their warm breeding grounds near the Equator to feeding areas in cooler waters near the poles.

Above: Bats spend long periods hibernating. They mate before winter, and the females store the sperm inside their body, only releasing it on the eggs as spring approaches.

They become dormant when their habitat becomes too dry. Most bury themselves in moist sand or under rocks, only becoming active again when rain brings the habitat back to life. Some aestivating frogs even grow a skin cocoon which traps moisture and keeps their bodies moist while they wait for the rains to return.

Migration

Another way of coping with bad conditions is to migrate. Migrations are not just random wanderings, but involve following a set route each year. In general they are two-way trips, with animals returning to where they started once conditions back home become acceptable again.

All sorts of animals migrate, from insects to whales, and there are many reasons for doing so. Most migrators are looking for supplies of food, or for a safe place to rear their young. For example, once their home territory becomes too crowded, young lemmings stampede over wide areas looking for new places to live, sometimes dying in the process. Herds of reindeer leave the barren tundra as winter approaches, and head for the relative warmth of the forest. Mountain goats act in a similar way: having spent the summer grazing in high alpine meadows, they descend below the treeline when winter snow begins to cover their pastures.

Other migrations are on a much grander scale, and in some cases an animal's whole life can be a continual migration. Wildebeest travel in huge

herds across southern Africa in search of fresh pastures. They follow age-old routes but may take a detour if grass is growing in unusual places. Among the greatest migrants are the giant whales, which travel thousands of miles from their breeding grounds in the tropics to their feeding grounds near the poles. The cool waters around the poles teem with plankton food, while the warmer tropical waters are a better place for giving birth.

Day length

How do animals know that it is time to hibernate or migrate? The answer lies in the changing number of hours of daylight as the seasons change.

All animals are sensitive to daylight, and use it to set their body clocks or circadian rhythms. These rhythms

affect all bodily processes, including the build-up to the breeding season. The hibernators begin to put on weight or store food as the days shorten, and many migrants start to get restless before setting off on their journey. However, not all migrations are controlled by the number of hours of daylight. Some migrators, such as wildebeest and lemmings, move because of other environmental factors, such as the lack of food caused by drought or overcrowding.

Below: Moose move across wide regions of mountains, rivers and even roads, migrating south each winter. They are known to migrate up to 196km (120 miles) from Alaska to northern Canada. The trigger for this migration is the arrival of large accumulations of snow in their feeding grounds, leading them to search for warmer browsing grounds.

INTRODUCED SPECIES

Centuries ago, as people started exploring and conquering new lands, many animals travelled with them. In fact, that's the only way many animals could travel such long distances, often crossing seas. Many introduced species then thrived in their new habitats, often at the expense of the native wildlife.

Perhaps one of the first introduced animals was the dingo. This Australian dog is now regarded as a species in its own right, but it was introduced to the region by the Aboriginals. These migrants had domesticated the ancestors of the dingo, probably from grey wolves, many years before.

Dingoes were one of the first placental mammals to come to Australia, which had previously been populated by marsupial mammals. In a chain of events that has been repeated many times since, the introduced dingoes became feral (living as wild animals), and were soon competing with native hunters for food. Not only did the introduced mammals begin to take over from the marsupial predators, such as Tasmanian devils and marsupial wolves, but they also wiped out many smaller marsupials which were unable to defend themselves against these ferocious foreign hunters. In fact, many more of the native Australian marsupials have now become extinct, or are in danger of doing so, though not just dingoes are responsible. As European settlers have arrived over the last couple of centuries, they have also introduced many new animals, including cats.

Domestic animals

Looking around the European countryside, you would be forgiven for thinking that cows, sheep and other farm animals are naturally occurring species. In fact, all come from distant parts of the world. Over the centuries, livestock animals have been selectively bred to develop desirable characteristics, such as lean meat or high milk production. Despite this, they can be traced back to ancestral species. For example, goats – a domestic breed of an Asian ibex – were introduced to North Africa about 3,000 years ago. These goats, with their voracious appetites, did well feeding on dry scrubland. In fact they did too well, and had soon stripped the plants, turning even more of North Africa into desert. Similarly, horses introduced to the Americas by European settlers had

Left: Dingoes are wild dogs that live in Australia. However, they did not evolve there; no large placental mammals did. They are actually the relatives of pet dogs that travelled with the first people to reach Australia.

Cows and sheep

European cows are believed to be descendants of a now-extinct species of oxen called the auroch, while modern sheep are descended from the mouflon. From their beginnings in the Middle East, new breeds were introduced to all corners of the world, where they had a huge effect on the native animals and wildlife.

Above: Cattle have been bred to look and behave very differently from their wild ancestors. Few breeds have horns, and they are generally docile animals. Some breeds produce a good supply of milk, while others are bred for their meat.

Below: Sheep were one of the first domestic animals. They are kept for their meat and sometimes milk, which is used for cheese. The thick coat or fleece that kept their ancestors warm on mountain slopes is now used to make woollen garments.

Above: While the ancestors of domestic horses have become extinct, horses have become wild in several parts of the world. Perhaps the most celebrated of these feral horses are the mustangs, which run free in the wild American West, after escaping from early European settlers.

a marked effect on that continent. Native people, as well as the settlers, began to use them to hunt bison on the plains, which eventually gave way to cattle. Groups of the introduced horses escaped from captivity and began to live wild. The feral American horses were called mustangs.

Rodent invaders

While many animals were introduced to new areas on purpose as livestock or pets, other animals hitched a lift. For example, some animals were more or less stowaways on ships, but only those that could fend for themselves at their new destination were successfully introduced. These species tended to be generalist feeders, and none was more successful than rodents, such as mice and rats. In fact the house mouse is the second most widespread mammal of all,

after humans. It lives almost everywhere that people do, except in the icy polar regions, although it is very likely that rodents did reach these places but then failed to thrive in the cold.

The black rat – also known as the ship rat – has spread right around the world from Asia over the last 2,000 years. On several occasions it has brought diseases with it, including bubonic plague, or the Black Death, which has killed millions of people. Another prolific travelling rodent is the brown rat which is thought to have

spread from Europe, and now exists everywhere except the poles.

Rodents are so successful because they will eat almost anything and can reproduce at a prolific rate. These two characteristics have meant that mice and rats have become pests wherever they breed.

Below: With their sharp and ever-growing teeth, rodents are very adaptable animals. Mice and rats have spread alongside humans, and wherever people go, these little gnawing beasts soon become established, breeding very quickly and spreading into new areas.

CONSERVING WILDLIFE

With so many species facing extinction, conservationists have their work cut out. Conservationists try to protect habitats and provide safe places for threatened animals to thrive, but the activities of ordinary people can often also have an adverse effect on the future of natural habitats.

People give many reasons why wildlife should be conserved. Some argue that if all the forests were cleared and the oceans polluted, the delicate balance of nature would be so ruined that Earth would not be able to support any life, including humans. Others suggest that if vulnerable species were allowed to die, the natural world would not be sufficiently diverse to cope with future changes in the environment. Another reason to save diversity is that we have not yet fully recorded it. Also, there are undoubtedly many as yet unknown species – especially of plants – which could be useful to humankind, for example in the field of medicine. But perhaps the strongest argument for the conservation of wildlife is that it would be totally irresponsible to let it disappear.

Habitat protection

Whatever the reasons, the best way to protect species in danger of being wiped out is to protect their habitats so that the complex communities of

plants and animals can continue to live. However, with the human population growing so rapidly, people are often forced to choose between promoting their own interests and protecting wildlife. Of course, people invariably put themselves first, which means that the conservationists have to employ a range of techniques in order to save wildlife.

In many countries it has now become illegal to hunt certain endangered animals, or to trade in any products made from their bodies. Whales, gorillas and elephants are protected in this way. Many governments and charitable organizations have also set up wildlife reserves, where the animals stand a good chance of thriving. The oldest

Below: One of the main causes of deforestation is people clearing the trees and burning them to make way for farmland. The ash makes good soil for a few years, but eventually the nutrients needed by the crops run out and so the farmers often begin to clear more forest.

Above: If logging is done properly, it can make enough money to pay to protect the rest of the rainforest. Only selected trees are cut down and they are removed without damaging younger growth. Forests can be used to grow crops, such as coffee and nuts, without cutting down all the trees.

protected areas are in North America and Europe, where it is illegal to ruin areas of forest wilderness and wetland. Consequently, these places have become wildlife havens. Other protected areas include semi-natural landscapes which double as beauty spots and tourist attractions. Although these areas often have to be extensively altered and managed to meet the needs of the visitors, most still support wildlife communities.

In the developing world, wildlife refuges are a newer phenomenon. Huge areas of Africa's savannahs are protected and populated with many amazing animals. However, the enormous size of these parks makes it very hard to protect the animals, especially elephants and rhinoceroses, from poachers.

Reintroduction

Large areas of tropical forests are now protected in countries such as Brazil and Costa Rica, but often conservation efforts come too late because many animals have either become rare or are completely absent after years of human

Zoo animals

Once zoos were places where exotic animals were merely put on display. Such establishments were increasingly regarded as cruel. Today, the world's best zoos are an integral part of conservation. Several animals, which are classified as extinct in the wild, can only be found in zoos where they are being bred. These breeding programmes are heavily controlled to make sure that closely related animals do not breed with each other. Later, individual animals may be sent around the world to mate in different zoos to avoid in-breeding.

Below: Many of the world's rarest species are kept in zoos, partly so that people can see them, since they are too rare to be spotted in the wild. Bears do well in zoos when they are provided with an enriched environment that includes food puzzles.

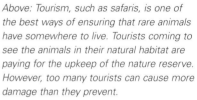

Above: Tourism, such as safaris, is one of the best ways of ensuring that rare animals have somewhere to live. Tourists coming to see the animals in their natural habitat are paying for the upkeep of the nature reserve. However, too many tourists can cause more damage than they prevent.

damage. However, several conservation programmes have reintroduced animals bred in zoos into the wild.

To reintroduce a group of zoo-bred animals successfully into the wild, conservationists need to know how the animal fits into the habitat and interacts with the other animals living there. In addition, for example when trying to reintroduce orang-utans to the forests of Borneo, people have to teach the young animals how to find food and fend for themselves.

Below: Breeding centres are an important way of increasing the number of rare animals. Most, such as this giant panda centre in China, are in the natural habitat. If the animals kept there are treated properly, they should be able to fend for themselves when released back into the wild.

Understanding habitats

A full understanding of how animals live in the wild is also vitally important when conservationists are working in a habitat that has been damaged by human activity. For example, in areas of rainforest which are being heavily logged, the trees are often divided into isolated islands of growth surrounded by cleared ground. This altered habitat is no good for monkeys, which need large areas of forest to swing through throughout the year. The solution is to plant strips of forest to connect the islands of untouched habitat, creating a continuous mass again.

Another example of beneficial human intervention involves protecting rare frogs in the process of migrating to a breeding pond. If their migration necessitates crossing a busy road, it is likely that many of them will be run over. Conservationists now dig little tunnels under the roads so that the frogs can travel in safety. Similar protection schemes have been set up for hedgehogs and ducks, to allow them safe passageways.

CLASSIFICATION

Scientists classify all living things into categories. Members of each category share features with each other – traits that set them apart from other animals. Over the years, a tree of categories and subcategories has been pieced together, showing how all living things seem to be related to each other.

Taxonomy, the scientific discipline of categorizing organisms, aims to classify and order the millions of animals on Earth so that we can better understand them and their relationship to each other. The Greek philosopher Aristotle was among the first people to do this for animals in the 4th century BC. In the 18th century, Swedish naturalist Carolus Linnaeus formulated the system that we use today.

By the end of the 17th century, naturalists had noticed that many animals seemed to have several close relatives that resembled one another. For example lions, lynxes and domestic cats all seemed more similar to each other than they did to dogs or horses. However, all of these animals shared common features that they did not share with frogs, slugs or wasps.

Linnaeus devised a way of classifying these observations. The system he set up – known as the Linnaean system – orders animals in a hierarchy of divisions. From the largest division to the smallest, this system is as follows: kingdom, phylum, class, order, family, genus, species.

Each species is given a two-word scientific name, derived from Latin and Greek. For example, *Panthera leo* is the scientific name of the lion. The first word is the genus name, while the second is the species name. Therefore *Panthera leo* means the "*leo*" species in the genus "*Panthera*". This system of two-word classification is known as binomial nomenclature.

Above: The bobcat (Felis rufus) *is an example of a small cat of the* Felidae *family, in the genus* Felis. *Other small cats include the pampas cat of South America and the jaguarundi of Central and South America.*

Below: The jaguar (Panthera onca) *belongs to the genus* Panthera, *the big cats, to which lions and tigers also belong. All cats are members of the order* Carnivora *(carnivores) within the class of* Mammalia *(mammals).*

Lions, lynxes and other genera of cats belong to the *Felidae* family. The *Felidae* are included in the order *Carnivora*, along with dogs and other similar predators. The *Carnivora*, in turn, belong to the class *Mammalia*, which also includes horses and all other mammals.

Mammals belong to the phylum *Chordata*, the major group which contains all vertebrates, including reptiles, amphibians, birds, fish and some other small animals called tunicates and lancelets. In their turn, *Chordata* belong to the kingdom *Animalia*, comprising around 31 living phyla, including *Mollusca*, which contains the slugs, and *Arthropoda*, which contains wasps and other insects.

Although we still use Linnaean grouping, modern taxonomy is worked out in very different ways from the ones Linnaeus used. Linnaeus and others after him classified animals by their outward appearance. Although they were generally correct when it came

Close relations

Cheetahs, caracals and ocelots all belong to the cat family *Felidae*, which also includes lions, tigers, wildcats, lynxes and jaguars. Although big cats can generally be distinguished from *Felis* by their size, there are exceptions. For example the cheetah is often classed as a big cat, but it is actually smaller than the cougar, a small cat. One of the main differences in the way the two groups behave is that big cats can roar but not purr continuously, while small cats are able to purr but not roar.

Above: The cheetah (Acinonyx jubatus) *differs from all other cats in possessing retractable claws without sheaths. This species is classed in a group of its own, but is often included within the group of big cats.*

Above: The caracal (Caracal caracal) *is included in the group of small cats (subfamily* Felinae), *but most scientists place the caracal in a genus of its own,* Caracal, *rather than in the main genus,* Felis.

Above: The ocelot (Felis pardalis) *is a medium-sized member of the* Felis *or small cat genus. Like many cats, this species has evolved a spotted coat to provide camouflage – unfortunately attractive to hunters.*

Distant relations

All vertebrates (backboned animals) including birds, reptiles and mammals such as seals and dolphins, are thought to have evolved from common fish ancestors that swam in the oceans some 400 million years ago. Later, one group of fish developed limb-like organs and came on to the land, where they slowly evolved into amphibians and later reptiles, which in turn gave rise to mammals. Later, seals and dolphins returned to the oceans and their limbs evolved into paddle-like flippers.

Above: Fish are an ancient group of aquatic animals that mainly propel themselves by thrashing their vertically aligned caudal fin, or tail, and steer using their fins.

Above: In seals, the four limbs have evolved into flippers that make highly effective paddles in water but are less useful on land, where seals are ungainly in their movements.

Above: Whales and dolphins never come on land, and their ancestors' hind limbs have all but disappeared. They resemble fish but the tail is horizontally – not vertically – aligned.

to the large divisions, this method was not foolproof. For example, some early scientists believed that whales and dolphins, with their fins and streamlined bodies, were types of fish and not mammals at all. Today, accurate classification of the various genera is achieved through a field of study called cladistics. This uses genetic analysis to check how animals are related by evolutionary change. So animals are grouped according to how they evolved, with each division sharing a common ancestor somewhere in the past. As the classification of living organims improves, so does our understanding of the evolution of life on Earth and our place within this process.

DIRECTORY OF ANIMALS

North and South America stretch from the barren ice fields of the Arctic
to the stormy coast of Cape Horn, which twists into the Southern Ocean.
A huge number of different animals make their home in the Americas
because the continents support a wide range of habitats. These include
the humid forests of the Amazon – the world's largest jungle; the searing
Atacama Desert – the world's driest place; and the rolling plains of the
American wild west. The Americas are home to many of the world's
record-breaking animals. The world's largest deer, the moose, is a
common resident of the conifer forests of Canada and the northern
United States. The largest land carnivore, the Kodiak bear, is found on
Alaska's Kodiak Island. The world's largest rodent, the capybara, lives in
the Amazon rainforest, and the largest snake in the world, the green
anaconda, is found in the wetlands across tropical South America.

Above from left: Caribou, grizzly bear, double-crested basilisk.

ALASKA

GREENLAND

ARCTIC CIRCLE

CANADA

UNITED STATES OF AMERICA

MEXICO

THE BAHAMAS

TURKS AND
CAICOS ISLANDS

CUBA

CAYMAN
ISLANDS

VIRGIN ISLANDS
SAINT KITTS AND NEVIS

DOMINICAN
REPUBLIC
HAITI

ANTIGUA AND BARBUDA

BELIZE

JAMAICA

PUERTO
RICO

GUADELOUPE
DOMINICA

GUATEMALA

HONDURAS

SAINT LUCIA

EL SALVADOR

NICARAGUA

SAINT VINCENT AND THE GRENADINES

BARBADOS

NETHERLANDS
ANTILLES

ARUBA

GRENADA

COSTA RICA

TRINIDAD AND TOBAGO

ATLANTIC OCEAN

PANAMA

VENEZUELA

GUYANA

FRENCH
GUIANA

GALAPAGOS
ISLANDS

COLOMBIA

SURINAM

ECUADOR

PERU

BRAZIL

BOLIVIA

PARAGUAY

SAN FÉLIX AND SAN AMBROSIO

CHILE

JUAN FERNANDEZ ISLANDS

URUGUAY

ARGENTINA

PACIFIC OCEAN

FALKLAND ISLANDS

SOUTH GEORGIA

SOUTH
SANDWICH
ISLANDS

SALAMANDERS

Salamanders and newts are amphibians with tails. All of them have legs, although a few species have lost a pair or have vestigial limbs. Like all amphibians, salamanders need a certain amount of water to reproduce. Some species are completely aquatic, while others live entirely on land. Many species are truly amphibious, spending the early part of their lives in water and living both on land and in water as adults.

Hellbender

Cryptobranchus alleganiensis

Hellbenders are among the largest salamanders in the Americas and one of three giant salamanders in the world. These monstrous and heavyset amphibians spend their entire lives on the beds of rivers and streams.

Hellbenders are nocturnal and spend the day sheltering under rocks. At night the animals become more active, but generally lurk in crevices while waiting for prey. Giant salamanders lose their gills as they change from larvae into adults. They absorb most of their oxygen through their wrinkly skin but will sometimes rise to the surface to take gulps of air into their small lungs.

Hellbenders breed in late summer. A male digs a hole under a rock and will only allow females that are still carrying eggs into his hole. Several females may lay their eggs in a single male's hole before he fertilizes them with a cloud of sperm. The male guards them for three months until the young hatch out.

Hellbenders have a dark, wrinkled skin that secretes toxic slime. These salamanders do not have gills, and the wrinkles in their skin increase the surface area of their bodies so that they can absorb more oxygen directly from the water.

Distribution: Eastern North America.
Habitat: Rivers and streams.
Food: Crayfishes, worms, insects, fish and snails.
Size: 30–74cm (12–30in).
Maturity: 2–3 years.
Breeding: 450 eggs laid in late summer.
Life span: 50 years.
Status: Not known.

Greater siren

Siren lacertina

Distribution: Eastern United States.
Habitat: Swamps, streams and lakes.
Food: Crayfishes, worms and snails.
Size: 50–90cm (20–36in).
Maturity: Not known.
Breeding: Eggs laid in spring.
Life span: 25 years.
Status: Not known.

The greater siren lives in the mud on the bottom of slow-flowing creeks and in swamps. Most salamanders change considerably as they mature into adults, but the adult body of a siren, with its external gills, long tail and single pair of legs, is very similar to the larval form. Greater sirens spend the day resting on the bottom. At night they drag themselves through the mud with their small legs or swim, with an S-shaped motion, through the murky water. These salamanders do not have teeth, but suck their prey through tough, horny lips. Greater sirens sometimes live in seasonal pools, which dry up in the summer. The salamanders survive these droughts by burying themselves in the moist sediment and coating their bodies with slimy mucus. Breeding takes place at night, under mud. It is thought that females lay single eggs on water plants and males follow the females around, fertilizing each egg soon after it is laid.

Greater sirens have very long bodies with feathery gills behind their heads and a single pair of legs. Their bodies are mottled to help them hide on the river bed. The salamander propels itself through the water, twisting its body into S-shaped curves.

Mudpuppy

Necturus maculosus

Mudpuppies and their close relatives, waterdogs, derive their names from the myth that they bark like dogs when handled. This confuses them with greater sirens – sirens being winged creatures from Greek mythology reputed to have lured sailors on to rocks with their song.

Mudpuppies spend their whole lives underwater. They do not change a great deal when they metamorphose from larvae into adults, retaining gills for breathing and long tails for swimming. The gills vary in size depending on how much oxygen there is dissolved in the water. They are very large in stagnant pools, where there is not very much oxygen available, and smaller in faster-running streams. Mudpuppies mate in the late autumn or early winter. The females' eggs are fertilized inside their bodies and they do not lay them until the spring. Before laying, the females make nests in hollows under rocks or logs on the beds of shallow streams. The eggs have sticky gelatinous coatings and stick together in layers.

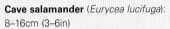

Mudpuppies have long, mottled bodies with two pairs of legs and a laterally flattened tail. There is a pair of feathery gills behind the head, but mudpuppies also have small lungs.

Distribution: Eastern United States.
Habitat: Muddy ponds, lakes and streams.
Food: Aquatic insects, crayfishes and fish.
Size: 29–49cm (12–20in).
Maturity: 4–6 years.
Breeding: Mate in autumn, eggs laid in spring.
Life span: Not known.
Status: Common.

Cave salamander (*Eurycea lucifuga*): 8–16cm (3–6in)
Cave salamanders are pale orange amphibians with black spots on their backs. They live in the mouths of caves in the Midwest region of the United States. Cave salamanders have long, grasping tails used for clinging to the rocky cave wall. They lack lungs, but still absorb all of their oxygen from the air through their skin and through the linings of their mouths. Because of this, they can only live in damp areas while on land and have long periods of inactivity.

Olympic torrent salamander
(*Rhyacotriton olympicus*): 9–12cm (3.5–5in)
The Olympic torrent salamander is one of four salamanders that live in the rocky streams of the Pacific Northwest in North America. They live in crystal-clear streams that tumble down mountainsides. The water is so rich in oxygen that the adult salamanders have only tiny lungs because they can easily get enough oxygen through their skins.

Coastal giant salamander (*Dicamptodon tenebrosus*): 17–34cm (6.5–13.5in)
Coastal giant salamanders are the largest land-living salamanders in the world. They have brown skins, mottled with black, which help them blend in with the floor of their woodland homes. They secrete foul-tasting chemicals through their skins as a defence when attacked by predators.

Three-toed amphiuma

Amphiuma tridactylum

With their slimy, cylindrical bodies, three-toed amphiumas and other species are also known as Congo eels. This is a misleading name because they are amphibians, not eels, which are fish. Also, they live only in North America, not in the African Congo.

Three-toed amphiumas are the longest salamanders in the Americas, although large hellbenders are probably heavier. Adult amphiumas have neither lungs nor gills. They have to take in oxygen directly through their skin. Three-toed amphiumas spend most of their lives in water, foraging at night and sheltering in streambed burrows during the day. However, during periods of heavy rainfall, the salamanders may make brief trips across areas of damp ground.

Three-toed amphiumas begin to mate in late winter. The females rub their snouts on males they want to attract. During mating males and females wrap around each other, while sperm is transferred into the female's body. The females guard their strings of eggs by coiling their bodies around them until they hatch, about 20 weeks later. By the time the eggs hatch, the water level may have dropped so the larvae have to wriggle over land to reach the water.

Three-toed amphiumas have very long, cylindrical bodies with laterally flattened tails. They have two pairs of tiny legs, each with three toes. The legs are too small to be of use for locomotion.

Distribution: Southern United States.
Habitat: Swamps and ponds.
Food: Worms and crayfishes.
Size: 0.5–1.1m (1.5–3.5ft).
Maturity: Not known.
Breeding: 200 eggs laid in spring.
Life span: Not known.
Status: Common.

Tiger salamander

Ambystoma tigrinum

Tiger salamanders start life in pools of water. While larvae, they have external gills and a fin down the middle of their tails. They feed on aquatic insects and even on other tiger salamander larvae. The larvae may stay in the water for several years before metamorphosing into the adult body form. The gills are then absorbed into the body, the tail fin is lost, and the skin becomes tougher and more resistant to desiccation. This change is thought to be catalyzed by warming of the water. Warm water can absorb more oxygen than cold water, which may trigger the change. However, the warming of the water may also suggest to the animal that the pool is in danger of drying up, indicating that it ought to leave.

   The metamorphosed salamander now looks like an adult, but it is still smaller than the mature animal. This subadult form is called an eft. The eft crawls out of the water and begins to forage on the leaf-strewn floor. It catches insects and other animals with flicks of its sticky tongue and then shakes them to death before chewing them up. After hibernating in burrows, mature adults make mass migrations in springtime to pools and mate over a period of two or three days.

Tiger salamanders may be dark green or grey with black markings, or yellow with black markings. Some specimens have yellow and black stripes, which make them even more reminiscent of their namesake.

Distribution: Central and south-eastern North America.
Habitat: Woodland.
Food: Worms and insects.
Size: 18–30cm (7–12in).
Maturity: Not known.
Breeding: Migrate to breeding pools in spring.
Life span: 20 years.
Status: Common.

Axolotl

Ambystoma mexicanum

Distribution: Lake Xochimilco, Mexico.
Habitat: Water.
Food: Aquatic insects.
Size: 10–20cm (4–8in).
Maturity: Not known.
Breeding: Not known.
Life span: 25 years.
Status: Vulnerable.

In the wild, most axolotls are black, but several color variants have been bred in captivity. In fact there are now more axolotls in captivity around the world than in the wild.

Axolotls are related to tiger salamanders and other species that spend their adult lives on land. However, axolotls never go through the change from the aquatic larval stage to the more robust, terrestrial adult form. Consequently, they spend their whole lives in water.

   Adult axolotls look almost identical to the larval stage. They have four legs – which are too small for walking on land – feathery gills behind their heads and long, finned tails used for swimming. Like the larvae, the adults feed on aquatic insects and other invertebrates, such as worms. The only major difference between the two forms is that the adults have sexual organs, so they can transfer sperm from the male to the female during mating.

   Axolotls do not metamorphose because their thyroid glands cannot produce the hormone necessary to bring about the change. Biologists have injected captive axolotls with the right hormone, and the salamanders then change into a land-living form similar to other species.

   Because they cannot naturally get out of the water, axolotls are confined to their aquatic habitat. They live in a single lake system in Mexico and are therefore vulnerable to pollution and exploitation, and are becoming increasingly rare.

Red-spotted newt

Notophthalmus viridescens

Red-spotted newts, also known as eastern newts, have a very complex lifecycle. Their lives begin in water, where they hatch out from eggs as aquatic larvae. The larvae then develop into land-living juveniles called efts. The eft stage has the body form of a normal adult salamander, with four legs and a long, grasping tail. However, it is unable to breed.

The eft spends up to four years living out of water in damp woodland and grassland. Red-spotted newts must return to ponds and streams to become mature. When they return to water, the maturing red-spotted newts actually redevelop many larval features, such as a deep tail – ideal for swimming – and a thinner skin, used for absorbing oxygen directly from the water.

The adults spend the spring and early summer breeding in bodies of fresh water. After the breeding season the adults leave the water and return to a life on land. Many newts return to the same ponds to breed each year. They may navigate by storing complex maps of their surroundings in their memories, or by using either the Earth's magnetic field or polarized light.

Red-spotted newts are so named due to the red and black markings along their backs. During the juvenile or eft stage, the newts are bright orange, but when they eventually reach maturity, they turn green with yellow undersides.

Distribution: Eastern United States.
Habitat: Damp land and fresh water.
Food: Tadpoles, insects, slugs, worms and other small invertebrates.
Size: 6.5–11.5cm (2.5–4.5in).
Maturity: 1–4 years.
Breeding: In water in spring.
Life span: Not known.
Status: Common.

Long-toed salamander
(*Ambystoma macrodactylum croceum*): 10–17cm (4–6.5in)
The long-toed salamander is found in the Pacific Northwest from northern Canada to British Columbia. It is a member of the mole salamander group, which includes tiger salamanders and axolotls. These amphibians are called mole salamanders because many, such as the long-toed salamander, live in burrows below damp ground. In spring, they breed in ponds.

Cayenne caecilian (*Typhlonectes compressicauda*): 30–60cm (12–24 in)
The Cayenne caecilian lives in the tributaries and lakes of the Amazon River basin system. Its long, dark body resembles that of an eel at first glance, but, like other caecilians, it has well-defined rings around its body. This species is completely aquatic: its tail is flattened into a paddle shape, and a small fin runs along the top of the body. Female Cayenne caecilians carry their developing eggs inside their bodies. Then they give birth to live young, which look like miniature adults.

Ringed caecilian

Siphonops annulatus

Caecilians are amphibians, but they belong to a separate group entirely from salamanders, newts, frogs and toads. All caecilians lack legs altogether and they resemble large worms. The ringed caecilian burrows through the soil of steamy tropical forests, wriggling its body in waves to push its way forward. They breathe with only one properly developed lung.

Ringed caecilians eat anything they come across while on the move, providing it is small enough. They use their heads like trowels to probe areas of soil and use their tentacles to pick up the scent of prey. The tentacles are closely linked to the nose, and detect chemicals in the soil. They can also detect the movements and faint electric currents produced by the muscles of prey animals.

Ringed caecilians mate when the soil is at its most damp, during the rainy seasons. The female's eggs are fertilized by the male when they are still inside her body. She lays the eggs in the soil. Unlike many species of caecilian, which have an aquatic larval stage, young ringed caecilians hatch looking like miniature adults.

Distribution: North-eastern South America.
Habitat: Tropical forest.
Food: Worms and insects.
Size: 20–40cm (8–16in).
Maturity: Not known.
Breeding: Rainy season.
Life span: 10 years.
Status: Common.

Ringed caecilians have blue, scaly bands around their bodies. Like all caecilians, ringed caecilians have a retractable tentacle on each side of the head, near to their nostrils.

FROGS AND TOADS

Frogs and toads form the largest group of amphibians, called Anura. Toads are better adapted to terrestrial habitats with thicker, warty skin to avoid desiccation, while frogs have thin, smooth skin that needs to be kept moist. Most species follow similar lifecycles. Tailed larvae called tadpoles hatch from eggs called spawn and develop in water before sprouting legs, losing their tails and emerging on to land.

Surinam toad

Pipa pipa

Distribution: Amazon Basin and northern South America.
Habitat: Muddy water.
Food: Small fish and aquatic invertebrates.
Size: 5–20cm (2–8in).
Maturity: Not known.
Breeding: Rainy season.
Life span: Not known.
Status: Common.

Surinam toads spend their whole lives in water. Their bodies are well adapted to a life foraging on the muddy bottoms of turbid streams and ponds. The fingers have a star of tentacles at their tips that makes them extremely sensitive feelers. The toads also have sense organs along their sides that can detect the motion in the water caused by other animals. Their eyes are positioned on the top of their heads so they can look up through the water to the see if danger is approaching from above the surface.

The toads mate during the wettest time of the year. A male grabs a female around the waist and the couple spin around in the water several times. The female releases her eggs and the male uses his hind feet to sweep them into the space between her back and his belly. His sperm fertilizes the eggs, which then embed themselves on the mother's back. The eggs develop into tiny toadlets after three or four months.

Surinam toads have flattened bodies with triangular heads and large webbed hind feet. Their forefeet have sensitive fingers for feeling around in murky waters. A female may bear up to 100 fertilized eggs on her back.

Darwin's frog

Rhinoderma darwinii

Darwin's frogs have small, slender bodies with pointed snouts and long fingers. The upper parts of their bodies are green or brown. The undersides are a dark brown or black. The long digits on the hind feet are webbed.

Darwin's frogs live in the steamy mountain forests of the Chilean Andes. Despite their natural habitat being damp and lush, Darwin's frogs do not have a typical aquatic life stage. Most frogs, toads and other amphibians spend at least part of their lives – generally the larval stage – living in water, but the water is in an unusual location in this case.

Darwin's frogs do not need this truly aquatic stage because of their unusual breeding system. The frogs breed at all times of the year. The male first attracts the female with a bell-like call. The female lays around 20 eggs in a suitably moist spot, and the male fertilizes the eggs and guards them for about 25 days until they hatch into tadpoles. He then scoops the tadpoles into his mouth. Males of a species related to Darwin's frogs carry the tadpoles to the nearest body of water, but Darwin's frogs keep the tadpoles in their mouths until they develop into froglets. The tadpoles develop in their father's vocal pouch for another 50 days before climbing out and becoming independent.

Distribution: Southern Andes, in southern Chile and Argentina.
Habitat: Shallow cold streams in mountain forests.
Food: Insects.
Size: 2.5–3cm (1–1.25in).
Maturity: Not known.
Breeding: All year.
Life span: Not known.
Status: Common.

South American bullfrog

Leptodactylus pentadactylus

Distribution: Central and northern South America, from Costa Rica to Brazil.
Habitat: Tropical forest near water.
Food: Insects and other invertebrates.
Size: 8–22cm (3–9in).
Maturity: Not known.
Breeding: Rainy season.
Life span: Not known.
Status: Common, although these frogs are hunted by humans in some areas and their hind legs eaten.

South American bullfrogs are not closely related to the bullfrogs of North America. They all have powerful bodies and large external eardrums, but the similarity ends there. The South American bullfrog is, in fact, more closely related to the horned frogs, which have pointed protuberances of skin above their eyes.

South American bullfrogs are mainly nocturnal. They shelter under logs or in burrows during the day and during periods when it is too dry to move around. These bullfrogs reportedly give a piercing scream when picked up. This probably serves to startle predators that might then drop them in fright.

South American bullfrogs breed in the wet season when the forest streams and ponds are swelled by the rains. The males whip up mucus into a blob of foam, using their hind legs, and attach it to a branch over a body of water. The females then choose a male's foam nest in which to lay their eggs. The tadpoles hatch out and fall into the water below.

The South American bullfrog is a large, robust animal with long limbs and widely spaced toes without webbing. The males have sharp spines on their forethumbs. During the mating season they use these weapons in fights with other males over the females.

Four-eyed frog

Physalaemus nattereri

Distribution: Southern Brazil and northern Argentina.
Habitat: Coastal forest.
Food: Insects.
Size: 3–4cm (1.25–1.5in).
Maturity: Not known.
Breeding: Rainy season.
Life span: Not known.
Status: Not known.

Four-eyed – or false-eyed – frogs live in the tropical forest near the Atlantic coast of South America. The adults spend their lives foraging on land. They breed after heavy rains have created ponds and puddles for the tadpoles to develop in. The male whips up a nest of mucus and foam, placing it near to a body of water. The female's eggs hatch into tadpoles, which wriggle or drop the short distance into the water.

While most frogs rely on poisons in the skin or alarm calls to ward off predators, four-eyed frogs have a different strategy. When threatened, they inflate their bodies so that they appear to be much larger. They then turn around and point their rump at the attacker. The eyespots on the frog's rump fool the predator into thinking that it is looking at the face of a much larger, and potentially dangerous, animal. To make matters worse, the frogs secrete a foul-smelling fluid. This fluid is released from a gland in the frog's groin.

Four-eyed frogs are so named because of the two black eyespots on their rumps.

Tailed frog (*Ascaphus truei*): 4cm (1.5in)
Tailed frogs lives in the clear mountain streams of the Cascade Range in the north-western United States and southern Canada. They have flattened heads with rounded snouts. The males have a short tail-like extension that contains their anus and sexual opening. The females lay short strings of eggs on rocks after mating with the males. The larvae develop into adults slowly, taking up to four years in colder parts of their range.

North American bullfrog (*Rana catesbeiana*): 9–20cm (3.5–8in)
The North American bullfrog is the largest frog in North America. It has a reputation for having a large appetite. It lives in lakes, ponds and slow-flowing streams, where it preys on small snakes, mammals and other frogs. During the summer breeding season, males defend territories and attract females by voicing deep croaks.

Greenhouse frog (*Eleutherodactylus planirostris*): 2–4cm (0.75–1.5in)
One of the world's smallest frogs, this species lives in Florida, Cuba and several other Caribbean islands. The greenhouse frog lives in woodland, where it uses suction discs on its fingers and toes to cling to flat surfaces, such as smooth bark, large leaves or, on occasion, greenhouse glass.

Marine toad

Bufo marinus

Distribution: Southern North America, Central and South America from Texas to Chile, and now introduced to other areas, including eastern Australia.
Habitat: Most land habitats, often near pools and swamps.
Food: Insects, particularly beetles, snakes, lizards and small mammals.
Size: 5–23cm (2–9in).
Maturity: 1 year.
Breeding: 2 clutches of between 8,000 and 35,000 eggs produced each year. Eggs hatch into tadpoles that become adult in 45–55 days.
Life span: 40 years.
Status: Common.

Marine toads are the largest toads in the world. They have several other common names, including giant toads and cane toads, their Australian name. Marine toads occur naturally from the southern USA southwards through Mexico to Chile. They were introduced to Queensland, Australia, in the 1930s to help control the pest beetles that were infesting sugar cane crops. However, the toads did not like living amongst the cane plants because there were few places to shelter during the day. So the toads spread out over the countryside, where they ate not beetle pests but small reptiles and mammals, some of which are now rare because of their predation. Today the toads are a serious pest.

Marine toads are extremely adaptable. They live in all sorts of habitats, eating just about anything they can get into their mouths. They protect themselves against attack using the toxin glands on their backs, which ooze a fluid that can kill many animals that ingest it. In small quantities, the toxin causes humans to hallucinate. Female cane toads produce several thousand eggs each year. They lay them in long strings, wrapped around water plants. The eggs are fertilized by the males after they have been laid.

Female marine toads are larger than the males. Both sexes have warty glands on their backs that squirt a milky toxin when squeezed.

Red-eyed tree frog

Agalychnis callidryas

Red-eyed tree frogs live high up in the trees in the rainforests of Central America. Their long legs allow them to reach for branches and then spread their body weight over a wide area when climbing through flimsy foliage. The discs on the tips of each toe act as suction cups, so the frogs can cling to flat surfaces, such as leaves.

Red-eyed tree frogs are nocturnal. Their large eyes gather as much light as possible so the frogs can see even on the darkest nights. During the day, the frogs rest on leaves. They tuck their bright legs under their bodies so that only their camouflaged, leaf-green upper sides are showing.

When it is time to breed, males gather on a branch above a pond. They call to the females by exciting clicking sounds. When a female arrives, a male climbs on to her back and she carries him down to the water. She takes in water and climbs back to the branch again, where she lays eggs on a leaf. The male fertilizes the eggs and they are left to develop into tadpoles. After hatching, the tadpoles fall into the water below.

Red-eyed tree frogs have long toes with rounded suction discs at their tips. Their bodies have a bright green upper side. These frogs have blue and white stripes on their flanks and yellow and red legs. The family of tree frogs, about 600 species strong, is found on all the continents except Antarctica.

Distribution: Caribbean coast of Central America.
Habitat: Tropical forests in the vicinity of streams.
Food: Insects.
Size: 4–7cm (1.5–2.75in).
Maturity: Not known.
Breeding: Summer.
Life span: Not known.
Status: Common.

Strawberry poison-dart frog

Dendrobates pumilio

Strawberry poison-dart frogs have bright red bodies with blue on their hind legs. The bright red and blue serve as a warning to predators that their skins are covered in a poison so deadly that just one lick is generally fatal. The poison-dart family has around 120 species.

Many frogs and toads secrete toxic chemicals on to their skins. Most are harmless to humans, but many make predators sick after they eat the frogs. However, the strawberry poison-dart frog and other closely related species have much more potent toxins. A single lick is enough to kill most predators.

The frogs earn their name from the fact that their skins are used by forest people to make poison for hunting darts. The toxins of some frogs are so strong that a single skin can produce enough poison to tip 50 darts. Hunters use them to kill monkeys and other forest animals.

The strawberry poison-dart frog is not always red. During the breeding season, males often change to brown, blue or green. Although they rarely climb trees when foraging, the females climb trees to lay their eggs in tree holes filled with water. The males then fertilize the eggs.

Distribution: Southern Central America.
Habitat: Tropical forest.
Food: Insects.
Size: 2–2.5cm (0.75–1in).
Maturity: Not known.
Breeding: Rainy season.
Life span: Not known.
Status: Common.

Mexican burrowing frog (*Rhinophrynus dorsalis*): 6–8cm (2.5–3.25in)
Mexican burrowing frogs have bloated, almost disc-shaped, black bodies and pointed snouts. They live in areas of Mexico and Central America with soft, sandy soils that are easy to burrow into. The frogs spend most of their time underground, only coming to the surface after heavy rain. They breed in temporary pools. The tadpoles are filter-feeders, straining tiny, floating plants and animals from the water.

Reticulated glass frog (*Centrolenella valerioi*): 2–5cm (0.75–2in)
This species of glass frog lives in the trees of mountain rainforests in South America. It is an expert climber, having slender legs with suction discs on its toes for clinging to the flat surfaces of leaves. The name "glass frog" refers to the species' translucent skin. The major bones and blood vessels are visible through it.

American toad (*Bufo americanus*): 5–9cm (2–3.5in)
The American toad is similar to its European cousin, the common toad, in that it has a brown, wart-covered body. It has large glands on its shoulder, which release a foul-tasting liquid on to the skin. American toads are found in most parts of eastern North America, from Hudson Bay to the Carolinas.

Paradoxical frog

Pseudis paradoxa

The paradoxical frog is aptly named. Most other frogs are considerably larger than their tadpoles. However, adult paradoxical frogs are smaller than their fully grown tadpoles, hence the paradox. Young paradoxical frogs stay in their larval tadpole stage for much longer than other species. They grow to 25cm (10in) long – four or five times the size of an adult. As they metamorphose into adults, the frogs therefore shrink in size, mainly by absorbing their tails back into their bodies.

Adult paradoxical frogs have bodies that are well adapted to a life in water. Their powerful hind limbs are webbed and are used as the main means of propulsion. Like many other aquatic frogs, this species has long fingers that are good for delving into the muddy beds of lakes and ponds. They stir up the mud to disturb prey animals and catch them in their mouths. The female lays her eggs in a floating foam nest, before the male fertilizes them.

Paradoxical frogs have slimy, dark green and brown bodies. They have very long hind legs with webbed feet. Their forefeet have two long toes.

Distribution: Central and eastern South America.
Habitat: Fresh water.
Food: Aquatic invertebrates.
Size: 5–7cm (2–2.75in).
Maturity: Not known.
Breeding: Rainy season.
Life span: Not known.
Status: Common.

TURTLES AND TORTOISES

Turtles and tortoises have lived on Earth for over 200 million years. They belong to a group of reptiles that have existed since the dinosaurs roamed the Earth. Their soft bodies are protected by shells called carapaces. There is no major difference between turtles and tortoises, however turtles (and terrapins) live in water, while tortoises tend to live on land.

Matamata

Chelys fimbriatus

Distribution: Amazon basin and northern South America.
Habitat: Beds of rivers and streams.
Food: Fish.
Size: 30–45cm (12–18in).
Maturity: Not known.
Breeding: 20–30 eggs laid.
Life span: 40 years.
Status: Common.

Matamatas live on the bottom of tropical lakes and rivers. They take breaths by poking their long snouts out of the water, so they can remain hidden below the water at all times. Their knobbly shells are often turned green and red by algae growing on them, helping them blend with rocky river beds. Matamatas wait in ambush for prey to swim near. They have small eyes, set on the sides of their flattened heads, which are useless for hunting in the murky waters, but they are sensitive to the water currents created by prey close by. The flaps of skin on their long necks aid their camouflage, and may also act as lures to attract fish. When the fish come within range matamatas strike with great speed. They suck the unsuspecting fish into their wide mouths. The suction is caused by rapid opening of the mouth, creating an area of low pressure inside. Matamatas have also been observed walking along river beds, herding fish into shallow water where they can be sucked up more easily.

Matamatas are unusual-looking turtles. They have triangular heads with long flexible snouts, which are used as snorkels to breathe air from above the surface.

Alligator snapping turtle

Macroclemys temminckii

Alligator snapping turtles are the largest freshwater turtles. During the day they are mainly ambush hunters, lying half-buried in mud on the river bed. While waiting for prey to approach, these turtles hold their large mouths open.

The turtle's tongue has a small projection on it, which becomes pink when engorged with blood. The turtle wiggles it as a lure to attract prey. Fish and other animals approach the lure, thinking they see a worm in the mud. However, they are heading for an almost certain death. As the prey swims into a turtle's mouth, it snaps shut. Small prey are swallowed whole, while the sharp, horny beak makes light work of larger prey, which may even be another species of turtle. The largest prey are held in the jaws, while the snapping turtle uses its forefeet to tear it apart. Male snapping turtles spend their whole lives in muddy rivers and lakes. Females, however, climb on to land in spring to lay eggs in holes dug into mud or sand.

Alligator snapping turtles have a tough carapace, covered in pointed, triangular knobbles. They have a large head with a sharp, horny beak.

Distribution: South-eastern United States, in the lower Mississippi River Valley.
Habitat: Beds of lakes and slow-flowing rivers.
Food: Fish and turtles.
Size: 40–80cm (16–32in).
Maturity: Not known.
Breeding: 10–50 eggs buried in mud.
Life span: 70 years.
Status: Vulnerable.

Galápagos tortoise

Geochelone elephantopus

Distribution: Galápagos Islands in the eastern Pacific.
Habitat: Varied on the islands, from moist forest to arid land.
Food: Plants.
Size: 1–1.4m (3.25–4.5ft).
Maturity: Not known.
Breeding: Several large eggs.
Life span: Over 100 years.
Status: Vulnerable in general; some sub-species are endangered.

Galápagos tortoises are the largest of all the testudines (tortoises and turtles). They are found only on the islands of the Galápagos Archipelago in the equatorial Pacific, off the coast of Ecuador. In general, the shell folds around the body like a saddle. However, the different subspecies located on various islands in the group have varying shell shapes. The general saddle shape allows the forefeet and neck to move more freely than for most tortoises.

Galápagos tortoises are plant-eaters. They use their long necks to reach up to bushes and shrubs, foraging for leaves with their toothless jaws. They also eat grass and even cacti. Their giant size is probably due to this diet. The Galápagos Islands are arid places and plants are not widely available. Plant food contains only small amounts of energy, and larger animals use energy more efficiently than small ones.

For most of the year, the tortoises live in small herds. During the breeding season, however, males defend territories. The dominant males are the ones that can lift their heads higher than other males. They hector passing females into mating with them. The females dig nest chambers and lay large spherical eggs.

These giant tortoises have different shell shapes depending on which island they live among the Galápagos Archipelago. Charles Darwin cited these shell differences to support his theory of natural selection.

Painted turtle (*Chrysemys picta*): 15–25cm (6–10in)
Painted turtles lives in fresh water across central North America, from British Columbia to Georgia. They have smooth shells with a red stripe down their backs, small limbs and yellow stripes along their necks. The turtles eat both plants and animals during the day and sleep on river beds at night.

Green turtle (*Chelonia mydas*): 1–1.2m (3.25–4ft)
Green turtles spend their whole lives at sea. They have flipper-like legs and smooth carapaces. They only visit land to lay eggs. The turtles gather in shallow water near deserted, sandy beaches across the world, including the Americas, where they mate. Then the females come out of the sea and lay their eggs in holes dug with their flippers in the sand. While at sea, the turtles eat jellyfish, sponges and molluscs.

Wood turtle (*Clemmys insculpta*): 14–19cm (5.5–7.5in)
Wood turtles wander between bodies of water in woodland and meadowland of eastern Canada and New England. They are omnivores, eating worms, slugs, tadpoles, insects and plants.

Stinkpot

Sternotherus odoratus

Stinkpots are so named because they release a nasty smelling musk from glands beneath their shells. This smell is meant to ward off predators, but the stinkpot will also give a painful bite if the musk does not do its job.

Stinkpots spend their lives in slow-flowing, shallow streams and muddy ponds and lakes. Their shells often have mats of microscopic algae growing on them. Stinkpots feed during both the day and night. They use the barbels on their chins to sense the movements of prey buried in the muddy streambeds. Like many other musk turtles, stinkpots have a toughened shelf attached to their upper jaws. The turtle uses this shelf to crush the shells of water snails and other prey.

Female stinkpots leave the water to lay their elongated eggs. They make nests under mats of decaying plant matter or under the stumps of trees. Stinkpots lay the smallest eggs of all turtles – only 1.5 × 2.5cm (0.5 × 1in).

Stinkpots have smooth, streamlined shells suitable for living in running water. They have sensitive fleshy projections, called barbels, on their chins.

Distribution: South-eastern United States.
Habitat: Shallow, muddy water.
Food: Insects, molluscs, plants and carrion.
Size: 8–13cm (3–5in).
Maturity: Not known.
Breeding: Eggs laid under tree stumps.
Life span: 54 years.
Status: Common.

LIZARDS

Lizards are reptiles, belonging to the same group as snakes. They are found all over the world except for Antarctica, especially in places that are too hot or dry for mammals to thrive. The main group of American lizards are the iguanas, which include the basilisks and anoles. More widespread lizards, such as geckos and skinks, are also found in the Americas, especially South America.

Rhinoceros iguana

Cyclura cornuta

Distribution: Island of Hispaniola and other Caribbean islands.
Habitat: Forest.
Food: Leaves and fruit.
Size: 1–1.2m (3.25–4ft).
Maturity: Not known.
Breeding: 2–20 eggs laid in burrow.
Life span: Not known.
Status: Vulnerable.

These large grey iguanas live on the island of Hispaniola – divided into the countries of Haiti and the Dominican Republic – as well as a few smaller islands in the Caribbean Sea. They are most active during the day and often bask in the sun to warm up their bodies. They walk slowly through the forest, browsing on leaves and fruit. Their teeth are very sharp and are ideal for cutting through tough leaves and other plant materials.

The rhinoceros iguana gets its name from the toughened scales on its snout that resemble small horns. Males have larger "horns" than females.

When threatened, rhinoceros iguanas will run away at high speed. They can only achieve these speeds over short distances. If cornered they will give a painful bite and thrash their tails, which are armoured with spiky scales.

Male rhinoceros iguanas maintain a hierarchy based on the size of their chin flaps, or dewlaps. The males frequently contest this social structure during the short summer breeding season. The top-ranked males control access to females. They attract females with elaborate displays involving bobs of their heads, press-ups and showing off their dewlaps. Females lay up to 20 eggs in burrows and guard them until they hatch three months later.

Gila monster

Heloderma suspectum

Gila monsters are one of only two poisonous lizards in the world. They produce venom in salivary glands in their lower jaws. The venom flows by capillary action along grooves in their teeth, giving the lizard a poisonous bite. The venom acts on the prey animal's nervous system, preventing the heart and lungs from working. For a healthy human, a bite from a gila monster will be very painful but not life-threatening.

Gila monsters are most active at night. They shelter from the heat of the day in rocky crevices or burrows abandoned by mammals. However, in northern parts of their range, the lizards are completely inactive for several months during the winter. Inactive individuals rely on fat stored in their tails to keep them alive when they cannot feed.

Distribution: South-west United States and northern Mexico.
Habitat: Desert.
Food: Small mammals and eggs.
Size: 35–50cm (14–20in).
Maturity: Not known.
Breeding: Eggs laid in summer.
Life span: 20 years.
Status: Vulnerable.

Gila monsters mate in springtime, and their copulation can last for over an hour. The eggs develop inside the females for about ten weeks. They then bury the eggs in areas that are often bathed in sunlight. The eggs incubate for up to ten months.

Gila monsters have long, robust bodies with short legs. Their bodies are covered in rounded, bead-like scales. Most of them are dark but some have blotches of pink, yellow or orange.

Green basilisk

Basiliscus plumifrons

Distribution: Central America.
Habitat: Forest.
Food: Insects, small vertebrates and fruit, often feeding in trees.
Size: 60–75cm (24–29in).
Maturity: Not known.
Breeding: 20 eggs per year.
Life span: 10 years.
Status: Common.

Green basilisks spend most of their time in trees. They have long fingers and toes that help them grasp branches as they scuttle about looking for food. However, they prefer to stay in trees that are close to water. When threatened by predators, such as birds of prey, the lizards dive into the water below. The crests on their backs and tails are used to propel the reptiles through the water to safety. Basilisks will also shimmy into soft sand to avoid a predator. They can close their nostrils to keep sand out. On land, they run at high speed on their hind legs. They do not have to stop running when they reach water because their hind feet have scaly fringes that spread the lizard's body weight, so they can run on the surface of water. This adaptive ability has earned them (along with other related species) the nickname of "Jesus Christ lizards".

Male basilisks control their territories very aggressively, chasing any other males away. A successful male will control an area that contains a number of females. He has sole mating rights to this harem.

Green basilisks are also known as plumed basilisks because they have crests on the backs of their heads, down their back and along their long tails.

Marine iguana (*Amblyrhynchus cristatus*): 50–100cm (20–40in)
Marine iguanas live on the coastline of the Galápagos Islands in the Pacific Ocean. They dive into seawater to feed on seaweed growing on rocks. Large adults can stay underwater for over an hour at a time. They spend the rest of the time basking in the sun on the rocky shoreline. They need to keep warm when on land because the islands are surrounded by cold ocean currents. Marine iguanas may gather in crowds containing thousands of individuals. They expel salt through glands in their noses. Their dark bodies become covered in pale salt crystals from the cold sea water.

Western banded gecko
(*Coleonyx variegatus*): 12–15cm (5–6in)
The western banded gecko is found in southern California, Arizona and northern Mexico. It is active at night, when it forages for insects and spiders. Its days are spent sheltering in rocky crevices. When trapped, this lizard gives high-pitched squeaks and, like many other gecko species, it may shed its tail to confuse a predator.

Green anole

Anolis carolinensis

Green anoles live in trees. Their long, thin legs are well adapted for leaping from branch to branch and perching on all but the flimsiest of branches. Thanks to the pads on the tips of their fingers and toes, anoles can grip on to just about any surface, including the fronds of palm trees.

Green anoles rest in dense cover at night. When not foraging in daylight, the lizards bask in the sun, generally on vertical surfaces, such as tree trunks or walls. Anoles can change their body color, although not to the same extent as chameleons. For example, they may darken their normally bright skin when resting in the shade so as not to attract attention.

Although both sexes have dewlaps, only the males use them for communication. They extend them to signal to rival males and mates. A male begins courtship by bobbing his head and displaying his dewlap to a female. He then walks towards her with his legs straightened. If she is receptive to his advances, she allows him to position his body next to hers. He then grasps the back of her neck with his mouth and holds her tail with his hind legs as they copulate. The female lays one egg at a time under moist leaf litter.

Both male and female green anoles have pink dewlaps – fans of skin beneath their throats. The lizards have very long tails – nearly twice the length of the rest of their bodies – and their elongated fingers and toes are tipped with pads.

Distribution: South-eastern United States.
Habitat: Woodland and shrubbery.
Food: Insects.
Size: 12–20cm (5–8in).
Maturity: Not known.
Breeding: Single eggs laid throughout breeding season.
Life span: Not known.
Status: Common.

Desert horned lizard

Phrynosoma platyrhinos

These small reptiles are sometimes referred to as horned toads, because of their rounded bodies. Despite living in dry areas, much of the desert horned lizard's range can get cold, especially at night. Its round body helps it warm up quickly in the morning sunshine. The lizard can eat huge quantities of ants, which it licks up with its long tongue.

The many spikes on their bodies serve two functions. They help to break up the profiles of the lizards so they can blend into the rocky terrain. If they are spotted by a sharp-eyed predator, the tough spikes make horned lizards a difficult and potentially painful meal.

Armed with these weapons, a horned lizard will not run when danger approaches. Instead, it will freeze to avoid giving its position away and rely on its camouflage to hide it. If this defensive strategy fails and it is scooped into the mouth of a predator, such as a coyote, the reptile has one final weapon. The horned lizard can ooze blood from membranes that haemorrhage around its eyeballs. The blood mixes with a foul-tasting chemical, causing the predator to release its grip on the lizard.

Distribution: South-western United States.
Habitat: Rocky desert.
Food: Ants.
Size: 7.5–13.5cm (3–5.5in).
Maturity: Not known.
Breeding: Eggs laid in spring.
Life span: Not known.
Status: Common.

The desert horned lizard has three pointed scales that form horns pointing backwards from the back of its head. There are smaller spikes on its back and along the tail.

Chuckwalla

Sauromalus obesus

Chuckwallas live in the Mojave Desert, one of the driest and hottest places in North America. Like all other living reptiles, chuckwallas are cold-blooded or exothermic. Their bodies do not make any heat of their own apart from that generated by muscle movement. The lizards rely on sunlight and the temperature of the air around them to warm up enough for daily activity. Chuckwallas only become fully active when their body temperature exceeds 38°C (100°F). The temperature of the Mojave Desert regularly exceeds this, but in other areas of their range, chuckwallas remain inside their rocky dens until the weather gets warm enough.

Chuckwallas are herbivores. They search through the rock-strewn desert for hardy plants that can survive the scorching conditions. With only a limited food supply, female chuckwallas may not be able to reproduce every year. Some females save energy by skipping a breeding season.

Chuckwallas have an unusual defence strategy. When they are under attack by a bird of prey or coyote, they scuttle into a tight crevice between rocks and inflate their lungs. These reptiles have loose folds of skin around their throats and flanks, which allow their bodies to swell up to a considerable size. This makes it difficult for a predator to extract the lizard from its hiding place.

Distribution: California, Arizona and northern Mexico.
Habitat: Desert.
Food: Fruit, leaves and flowers.
Size: 28–42cm (11–16in).
Maturity: Not known.
Breeding: Females breed every 1–2 years.
Life span: Not known.
Status: Common.

Chuckwallas have powerful limbs and thick tails. The males have completely black heads, while the females have yellow and orange patches on black.

Black tegu

Tupinambis teguixin

Black tegus are among the largest lizards in South America. They live in forest clearings, often close to river banks. They are good swimmers and can also walk on their hind legs.

Black tegus find their prey by tasting the air with their forked tongues. The lizards flick their tongues in the air and then place each prong into a double olfactory organ at the top of their mouths. The organ is used to analyze the air for any chemicals contained within it.

When they are attacked, these large lizards scratch and bite predators and often use their muscular tails as clubs. Black tegus communicate with each other with loud snorting noises, which carry through the forest. The males have paired sexual organs called hemipenes. During mating, the male grabs the female's neck and twists his body under hers. He then inserts his hemipenes into her sexual opening, which is called a cloaca. Females may lay eggs in burrows, but typically they dig into the bases of termite mounds because termites keep these nests at a constant temperature.

The black tegu has a thick tail and powerful limbs. Its scales are small and give the body a glossy appearance. Although it is not related, the black tegu resembles the monitor lizards.

Distribution: South America from Panama to Argentina.
Habitat: Tropical forest.
Food: Insects, lizards, small birds and mammals.
Size: 80–110cm (31–43in).
Maturity: Not known.
Breeding: Eggs laid in termite mounds.
Life span: Not known.
Status: Common.

Pink tailed skink (*Eumeces lagunensis*): 16–20cm (6–8in)
This skink lives in Baja California. When young, the lizard's long tail is bright pink, but it fades with age. The bright tail probably serves as a diversion to predators, which are more likely to strike at the bright tail than the camouflaged head. Many lizards can live without their tails, and skinks have special bones in their tails that are designed to break when the tail is attacked. In most cases a lost tail will grow back again in a matter of weeks.

Middle American night lizard
(*Lepidophyma flavimaculatum*): 7–12cm (3–5in)
Female Middle American night lizards produce young without having to mate – parthenogenetically. Some populations of these lizards contain no males at all. The lizards do not lay eggs, but keep them inside their bodies until they hatch. The young are then born live from the mother.

Crocodile skink (*Tribolonotus gracilis*): 15–20cm (6–8in)
This armoured lizard lives on the island of Hispaniola in the Caribbean. Its triangular head is protected by a bony shield and it has four rows of scaly spines running along its back and tail, making it look somewhat like a crocodile. Crocodile skinks live in pairs or threesomes, consisting of a male and one or two females. The females lay single eggs in leaf litter.

Ajolote

Bipes biporus

Ajolotes are not snakes or lizards, but amphisbaenians, belonging to a small group of reptiles, which are sometimes called worm lizards. These reptiles spend their whole lives burrowing through soft soil, feeding on subterranean prey.

The ajolote is also called the mole lizard or "little lizard with ears". Early observers must have mistaken the ajolote's tiny but powerful forelimbs for ears. This is an easy mistake to make, since the forelegs are so near to the head. Ajolotes use their clawed forefeet to dig tunnels. They also wriggle through the soil, using their blunt heads to push earth aside.

Unlike other amphisbaenians, ajolotes sometimes come to the surface, generally after heavy rains. Above ground they are ambush hunters, lying in wait for lizards and other small animals. They use their forelimbs to drag their long bodies across the ground to grab their prey, which they generally drag underground to be eaten in safety.

Distribution: Baja California.
Habitat: Burrows.
Food: Worms and insects.
Size: 17–24cm (6–9.5in).
Maturity: Not known.
Breeding: Lays eggs.
Life span: 1–2 years.
Status: Common.

Ajolotes have long bodies with scales arranged in rings. They are unique among worm lizards because they still have forelimbs.

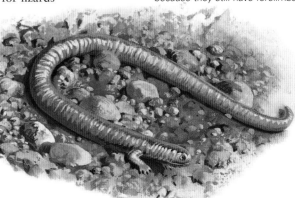

CROCODILIANS

Crocodilians are an ancient group of reptiles. They include crocodiles, alligators, caimans and gharials. They all live in or near water. Most crocodilians in North and South America belong to the alligator and caiman group. In fact only one type of alligator – the Chinese alligator – lives outside of the Americas. American crocodiles tend to be larger than their alligator and caiman cousins, and most are very rare.

American alligator

Alligator mississippiensis

Distribution: South-eastern United States.
Habitat: Swamps and rivers.
Food: Birds, fish and mammals.
Size: 2.8–5m (9.25–16.5ft).
Maturity: 5–10 years.
Breeding: Eggs laid in nest of mud and vegetation during spring.
Life span: 40 years.
Status: Common.

Young American alligators feed on insects, small fish and frogs. As they get bigger, they begin to take larger prey, such as turtles and water birds. Adults feed on land as well as in water. They are opportunistic feeders, attacking anything that comes within reach. They even leap up to snatch birds perching on low branches.

During cold weather, American alligators become dormant in burrows dug into mud banks. In dry periods, they will travel long distances to find water, sometimes ending up in swimming pools.

During courtship the female makes a mound of vegetation and mud above the high waterline and lays her eggs in a hole in the top. When she hears calling from hatchlings, she breaks open the nest and carries her young to the water. The young stay with the mother for about a year.

Unlike other crocodilians, the fourth teeth on either side of the lower jaw are not visible in alligators, fitting into sockets in their upper jaws.

Black caiman

Melanosuchus niger

Black caimans are the largest alligators in the Americas. When young they rely heavily on aquatic crustaceans for food, such as crabs and crayfish. They eat fish, such as catfish and piranhas, and often take large rodents called capybaras that live along the banks of rivers. At night, adults may hunt on land, taking advantage of their excellent hearing and sense of sight to track large animals, which may include livestock and even humans.

Breeding takes place during the dry season, presumably to reduce the chance that the eggs become submerged while they incubate. The females build nest mounds that are about 1.5m (5ft) high. The nest mounds are built in a variety of places, some concealed, others in the open. Each female digs a conical hole in her nest and lays 30–65 eggs in the top. They hatch about three months later, at the beginning of the wet season.

Distribution: Northern South America.
Habitat: Rivers and flooded forests.
Food: Fish, capybaras and other aquatic vertebrates.
Size: 4–6m (13.25–20ft).
Maturity: 5–10 years.
Breeding: Eggs laid in nest during dry season.
Life span: 40 years.
Status: Lower risk.

As their name suggests, black caimans have dark bodies. They have grey-brown bands on their lower jaws. The young have yellow or white bands on their flanks, which fade as they age.

Spectacled caiman

Caiman crocodilus

Distribution: Central America and northern South America.
Habitat: Areas of still water.
Food: Fish, wild pigs and water birds.
Size: 1.5–2.5m (5–8.25ft).
Maturity: 4–7 years.
Breeding: Eggs laid in nests during wet season.
Life span: 40 years.
Status: Common.

Spectacled caimans live in a wide range of habitats from rivers to coastal wetlands. They prefer stiller waters than the black caimans that share parts of their range, and consequently they have taken up residence in many reservoirs.

Spectacled caimans rarely come out of the water, only attacking land animals when they come to the water's edge. They spend the day floating on the surface and hunt mainly at night. During periods of drought, they aestivate – enter a period of dormancy to avoid desiccation – in cool burrows dug deep into mud.

The breeding season coincides with the wet season in May and June. The dominant males get the best territories and attract most females. Females lay eggs in mounds of vegetation that they build on banks or rafts of plants. Several females may lay eggs in a single nest, which they guard together. The young live in large groups called crèches.

Spectacled caimans are smaller than most crocodilians. They get their name from ridges of bone located on their snouts between their eyes. The ridges appear to join the eyes together, looking similar to the frames of a large pair of glasses.

Orinoco crocodile (*Crocodylus intermedius*): 4–7m (13.25–23ft)
Orinoco crocodiles live in the middle and lower parts of the Orinoco River. The crocodiles also live in the wetland created when the river floods the grassland on either side. The crocodiles eat birds and land animals as well as fish. The females lay their eggs in holes dug into sandbars during periods of low water – in January and February. The eggs hatch as the rains arrive about 70 days later. The young are protected by their mothers for up to three years.

Cuvier's dwarf caiman (*Paleosuchus palpebrosus*): 1.2–1.6m (4–5.25ft)
This small caiman lives in South America, from Paraguay to Colombia. It inhabits freshwater rivers and flooded forests. It spends the day in burrows and feeds on small aquatic and terrestrial animals. Its short, curved teeth are particularly good at crushing shellfish.

Cuban crocodile (*Crocodylus rhombifer*): 3.5m (11.5ft)
This rare crocodile lives in the swamps of Cuba. It mainly eats fish and turtles, but will take small mammals if the opportunity arises. Its broad, back teeth can crush turtle shells. Cuban crocodiles are very agile on land. Some scientists believe that they may have once preyed upon giant ground sloths, which are now extinct.

American crocodile

Crocodylus acutus

American crocodiles live in fresh water, such as rivers and lakes, but will venture out into coastal waters, especially near estuaries and in lagoons, where the water is brackish. The crocodiles cope with the salty water by taking long drinks of fresh water when possible and removing salt from the body through glands on their faces – secreting crocodile tears in the process.

During periods of drought, the crocodiles burrow into mud and do not feed until the water returns. Feeding takes place at night and they may come on to land to prey on livestock. They have also been known to attack humans.

Most females lay their eggs in holes dug in the ground, but they may build nest mounds in areas where the soil is likely to become waterlogged and thus chill the incubating eggs. Nesting takes place in the dry season. Between 30 and 60 eggs are laid, which hatch three months later as the rainy season begins. The mother guards her nest until the hatchlings have dispersed.

Unlike those of the alligators, the American crocodile's long fourth teeth are visible, sticking out of the lower jaw when the mouth is closed. American crocodiles often exceed the size of other crocodilians in North and South America.

Distribution: Southern Florida, Mexico, Central America and northern South America.
Habitat: Rivers and brackish water.
Food: Fish, turtles and birds.
Size: 4–5m (13.25–16.5ft).
Maturity: 5–10 years.
Breeding: Eggs laid in dry season.
Life span: 40 years.
Status: Vulnerable.

SNAKES

Snakes are legless reptiles that have adapted to living in most habitats on Earth. They are not found in the polar regions or high up in mountains. In the oceans, they are found in some tropical seas. All snakes are hunters of live prey particular to their chosen habitat. Familiar snakes of the Americas include the rattlesnakes and the green anaconda.

Green anaconda

Eunectes murinus

Distribution: Northern South America.
Habitat: Wetlands and flooded forests.
Food: Birds, caimans, deer and capybaras.
Size: 6–10m (20–33ft); 250kg (550lb).
Maturity: 6 years.
Breeding: 4–80 young born from mother.
Life span: 25 years.
Status: Common.

Green anacondas are the world's heaviest snakes, if not the longest. They are not venomous snakes but kill by constriction, squeezing their prey in coils of their massive bodies.

Green anacondas spend most of their time in shallow water, being most common in open wetlands. Their eyes and nostrils are positioned on top of their heads, so that they can lie hidden underwater with only their heads at the surface. Anacondas are ambush predators: they wait for prey to come to the water's edge to drink, then they strike with lightning speed. Their bodies are powerful enough to squeeze the life out of a horse or a fully grown black caiman. Anacondas can kill humans, but only occasionally do so.

Male anacondas have claw-like spurs on their lower bodies, which they use to stimulate females. A single female may be tangled up with several males during mating. Like other boas, anacondas do not lay eggs but give birth to live young.

Compared to their huge bodies, green anacondas have small heads. Their bodies are covered in smooth olive scales and they have black ovals on their backs. The males are smaller than the females. The young, born live from the mother, measure about 66cm (26in) at birth.

Emerald tree boa

Corallus caninus

Emerald tree boas spend their entire lives away from the ground, gripping tree branches with their coils. The snake's bright leaf-green body has flashes of white running across its back that help it blend in with the forest foliage. This camouflage keeps the snake safe from forest raptors, such as owls and eagles.

Tree boas hang from sturdy branches and wait for small birds to fly by or small mammals to pass beneath them. The snake's eyes have vertical pupils. Just like small cats, this makes them better at sensing the movements of small prey in the gloom of the forest. Tree boas also have heat-sensitive pits on their snouts, which allow them to detect the body heat of prey moving near them. The snake waits, ready to pounce, with its upper body in an S-shape. When a prey animal comes close enough, the tree boa lunges forward and grabs it in its mouth. The snake's backward-curved teeth stop victims from struggling free.

During mating, the male entwines his tail with the female's. The female gives birth to between 3 and 15 young. The young snakes are red or orange for the first year of their lives.

Distribution: Northern South America.
Habitat: Rainforest.
Food: Birds and mammals.
Size: 1.5–2m (5–6.5ft).
Maturity: Not known.
Breeding: Young born live.
Life span: 40 years.
Status: Common.

The emerald tree boa has a long, slender body with a prehensile tail that is used for gripping branches. The snake's camouflaged body is reinforced down each side, like a girder, so it is powerful enough to reach across open spaces between branches.

Rosy boa (*Charina trivirgata*): 60–110cm (24–43in)
Rosy boas live in the deserts of the south-western United States and northern Mexico. Compared to other boas, their heads are less broad in proportion to their bodies. They have smooth scales that form red and cream stripes along their bodies. Rosy boas burrow under rocks and slither in crevices looking for prey.

Boa constrictor (*Boa constrictor*): 1–4m (40–160in)
Boa constrictors occur in a range of colors and patterns. Their coloration often depends on where they live, since they are equally at home in the branches of rainforest and among the grasses of open savannah. These boas have also become a common sight in many South American cities.

Red pipesnake (*Anilius scytale*): 70–90cm (28–35in)
This unusual snake uses its cylindrical body to tunnel underground, where it preys upon other subterranean animals, such as rodents and other snakes. Although it is hidden from view much of the time, the red pipesnake has striking red and black bands on its body. This pattern mimics the colors of highly venomous coral snakes, so predators stay away, fearing they may receive a deadly bite.

Kingsnake

Lampropeltis getulus

Kingsnakes are constrictors, killing their prey by squeezing them to death. They are very active hunters, slithering into rodent burrows and climbing through bushes to catch their diverse prey. Kingsnakes appear to be immune to the venom of other poisonous snakes, which they include on their menu. Kingsnakes are able swimmers and patrol riverbanks in search of frogs and small aquatic mammals.

As well as threatening many animals, kingsnakes also have several enemies of their own, from large birds of prey to raccoons and other carnivores. If cornered, a kingsnake will try to bite its attacker. If captured, the kingsnake's final defence is to smear its captor with foul-smelling faeces.

Kingsnakes live in a variety of climates and may be active during the day or night. They hibernate during cold periods, which may last for several months at the northern extent of their range.

The males bite their mates on the backs of their necks to restrain them during mating. About 12 eggs are laid under rotting vegetation.

Kingsnakes have different colored bodies in different parts of their range. For example, Mexican kingsnakes are black, while those found in the deserts of Arizona have yellow bodies with black spots.

Distribution: Southern United States and northern Mexico.
Habitat: All land habitats.
Food: Birds, lizards, frogs and other snakes.
Size: 1–2m (3.25–6.5ft).
Maturity: Not known.
Breeding: Eggs laid in rotting vegetation.
Life span: 25 years.
Status: Common.

Texas thread snake

Leptotyphlops dulcis

The Texas thread snake, or Texas blind snake as it is also known, spends most of its life burrowing through the ground. Its body is well adapted for this lifestyle, being equipped with smooth scales and a blunt head for shoving earth out of the way.

The snakes feed on worms and other invertebrates that they come across, using their keen sense of smell to locate them. They also tunnel into the nests of ants and termites. When they enter a nest, the snakes begin to release the same chemical pheromones used by the insects themselves. This fools the normally aggressive insects into thinking the snakes belong there. The invading reptiles are free to slither about and feast on insect eggs and larvae. Texas thread snakes will come to the surface at night, especially after heavy rains, when the soil beneath is waterlogged.

After mating, the female lays only a handful of eggs and stays close to them while they incubate, often coiling around them. The thread snake family number 80 species in all.

The long, thin body of the Texas thread snake is covered in smooth, silvery scales. Even the snake's eyes are covered in thin scales.

Distribution: Southern United States and northern Mexico.
Habitat: Underground.
Food: Insects and spiders.
Size: 15–27cm (6–11in).
Maturity: Not known.
Breeding: Female incubates eggs in coils.
Life span: Not known.
Status: Common.

Sidewinder

Crotalus cerastes

Sidewinders are named for their unusual method of locomotion. The snakes move sideways across loose ground, such as sand. Many snakes that live in similar habitats also "sidewind". This is a form of locomotion that involves a wave-like undulation of the snake's body, so that only two points are in contact with the ground at any given moment. The snake progresses in a sideways direction across the ground compared to the orientation of the body, leaving parallel S-shaped tracks in the sand.

Sidewinders are rattlesnakes. They feed on small desert animals during the cool of night, using heat-sensitive pits below their eyes to detect the body heat of their prey. These snakes lie in wait for prey under the cover of small desert shrubs. Sidewinders strike with lightning speed, injecting venom supplied by glands in their upper jaws, which flows through their hollow fangs. The prey may manage to escape a short distance before being overcome by the sidewinder's venom. However, the snakes can soon locate the corpses of prey with their heat-sensitive pits.

Distribution: South-western United States and north-western Mexico.
Habitat: Desert.
Food: Lizards and rodents.
Size: 45–80cm (18–32in).
Maturity: Not known.
Breeding: Live young born in late summer.
Life span: Not known.
Status: Common.

Sidewinders have wide bodies so that they do not sink into sand. Their tails are tipped with rattles that increase in length as the snakes age. Their heads are flattened and triangular.

Western diamondback rattlesnake

Crotalus atrox

Distribution: Southern United States and northern Mexico.
Habitat: Grassland and rocky country.
Food: Vertebrate prey including small mammals, birds and lizards.
Size: 2m (6.5ft).
Maturity: 3–4 years.
Breeding: Young born live.
Life span: Not known.
Status: Lower risk.

Western diamondbacks are the largest and most venomous rattlesnakes in North America. The snake's rattle comprises dried segments – or buttons – of skin attached to the tail. The rattle is used to warn predators that the snake gives a poisonous bite. Although it will readily defend itself when cornered, the diamondback would prefer to conserve venom, and enemies, including humans, soon learn to associate the rattle with danger.

Like all rattlesnakes, diamondbacks are not born with a rattle. Instead they begin with just a single button, which soon dries into a tough husk. Each time the snake moults its skin, a new button is left behind by the old skin. The rattle grows in this way until it contains around ten buttons that give the characteristic noise when shaken.

Western diamondbacks have very potent venom. They kill more people each year than any other North American snake, although this number rarely reaches double figures. The venom is extremely effective at tackling prey. It can kill even large prey, such as hares, in seconds. Like other rattlesnakes, diamondbacks can sense body heat using pits on their faces.

Diamondbacks are so named because of the brown diamonds, bordered with cream scales, seen along their backs.

Western coral snake

Micruroides euryxanthus

Distribution: South-western United States and northern Mexico.
Habitat: Desert.
Food: Snakes and lizards.
Size: 60–90cm (24–35in).
Maturity: 1–2 years.
Breeding: Lay eggs in summer.
Life span: Not known.
Status: Common.

As they are relatives of cobras and mambas, western coral snakes have a similarly potent venom. Their bright bands serve as a warning to potential predators that the snakes are very dangerous. The venom will kill most small animals. One in ten humans who leave a bite untreated are overcome by the toxins.

Coral snakes spend a lot of time underground. Their thin, cylindrical bodies covered in smooth scales are ideal for this tunnelling lifestyle, and their rounded heads are used for burrowing through soft soil. However, the snakes also venture into the burrows of other animals to seek out resting snakes and lizards, their main source of food. They rarely come to the surface during the day, but may emerge to hunt in the cool of night. Western coral snakes mate in early summer and the females produce eggs about a month later. They are the only venomous snakes in the Americas to lay eggs rather than bear live young.

The western coral snake, with its brightly banded body, is typical among American coral snakes. Many non-venomous snakes mimic the colors of coral snakes to deter predators.

Milksnake (*Lampropeltis triangulum*): 0.4–2m (1.3–6.5ft)
Milksnakes are found ranging from Colombia to southern Canada. They are not venomous, but are among several species that mimic the red, yellow and black banded coloration of coral snakes. They are nocturnal hunters, preying upon small rodents and amphibians. Females lay eggs in nests constructed under rocks, in tree stumps and other secluded spots.

Corn snake (*Elaphe guttata*): 1–1.8m (3.25–6ft)
Corn snakes belong to the rat snake group. They live in the south-eastern United States and are found in several color forms. Corn snakes hunt on the ground, up trees and among rocks. The are not venomous but will rise up when threatened as if to strike. They kill their prey – mainly small rodents – by constriction.

Common lancehead (*Bothrops atrox*): 1–1.5m (3.25–5ft)
Lanceheads, or fer-de-lance snakes, are pit vipers. Like rattlesnakes, they have heat-sensitive pits on their faces. They are named for their arrow-shaped heads, typical among vipers. Lanceheads are among the most dangerous snakes in South America. Others have more potent venom, but lanceheads are often found living alongside people, feeding on rats and other rodents. Like many other vipers, male lanceheads tussle with each other over females.

Common garter snake

Thamnophis sirtalis

The common garter snake has one of the widest ranges of any North American snake. At one extreme, the common garter lives on the southern shores of Hudson Bay in eastern Canada and survives the long and icy subarctic winters. At the other end of its range, it lives in the humid, subtropical swamps of Florida.

Garter snakes are closely associated with water. They are able swimmers but feed on animals found in and out of water. Garter snakes are active hunters and generally have to pursue their victims. They seek out meals by poking their small heads into nooks and crannies and flushing out prey. Their long bodies allow them to move with great speed, and their large eyes are suited to tracking prey on the move.

Garter snakes hibernate in burrows, and many snakes may crowd into a suitable hole. Mating takes place soon after hibernation. In northern regions with short summers, the pressure to mate quickly is very strong, while in the south the snakes have a longer breeding season.

Distribution: Southern Canada to Florida.
Habitat: Close to water.
Food: Worms, fish and amphibians.
Size: 65–130cm (25–51in).
Maturity: Not known.
Breeding: Mate after hibernation.
Life span: Not known.
Status: Common.

Common garter snakes have long bodies and small heads. They are found in a variety of colors, which provide camouflage in different habitats.

CATS

Cats belong to the Felidae *family of mammals. They fall into two main groups. The* Panthera *genus contains the big cats, such as lions and tigers, while* Felis *comprises the small cats, including the domestic cat. Most American cats belong to the second group, with the jaguar being the only big cat found on both continents. Today most are rarely seen, and some are threatened with extinction.*

Cougar

Felis concolor

Extremely strong and agile, cougar adults are able to leap over 5m (16.5ft) into the air. Once they make a kill, their victims are dragged into secluded places and eaten over several days.

Cougars are also known as pumas, panthers or mountain lions, and have the most widespread distribution of any American species. They live in nearly all habitats, from mountainsides of the Canadian Rockies to the jungles of the Amazon and the swamps of Florida.

The cougar is the largest of the small cats in America, males reaching over 2m (6.5ft) long. They patrol large territories, moving during both day and night and taking shelter in caves and thickets. A cougar's preferred food is large deer, such as mule deer or elk. They stalk their prey before bringing it down with a single bite to the throat, or ambush it from a high vantage point. Cougars live alone, and mark their territories with their scent and by scraping visual signals in the soil and on trees.

Distribution: North, Central and South America from southern Canada to Cape Horn.
Habitat: Any terrain with enough cover.
Food: Deer, beavers, raccoons and hares.
Size: 1–2m (3.25–6.5ft); 60–100kg (135–220lb).
Maturity: 3 years.
Breeding: Every 2 years; litters of 3 or 4 cubs.
Life span: 20 years.
Status: Some subspecies endangered.

Margay

Felis wiedii

Distribution: Central America and Amazon Basin.
Habitat: Tropical forest.
Food: Birds, eggs, lizards, frogs, insects and fruit.
Size: 46–79cm (18–31in); 2.5–4kg (5.5–9lb).
Maturity: 1 year.
Breeding: Single cub or twins born once a year.
Life span: 10 years.
Status: Endangered.

Margays are small cats that live in the lush forests of Central and South America. These small, slender cats spend nearly all of their lives in the treetops, rarely touching ground. They are active at night, searching through the branches for food, which ranges from small tree-dwelling mammals, such as marmosets, to insects and fruit.

Margays are very acrobatic climbers. They use their long tails to help them balance, and their broad, padded feet give them a good grip on flimsy branches. Margays are unique among cats because they can twist their hind feet right round so they face backwards.

Like most cats, margays live alone, defending large territories from intruders. They do, of course, pair up briefly with mates for breeding, but the males leave the females before litters are born. Breeding takes place throughout the year and most litters have one or perhaps two cubs.

Margays can climb down tree trunks headfirst like squirrels, or hang upside down with the claws on their reversed hind feet embedded in tree bark.

Jaguar

Panthera onca

The jaguar is the only big cat in the Americas. It is smaller in length than the cougar, but much bulkier and heavier. Jaguars are usually a tawny yellow with dark rings, but they can also be black.

Jaguars prefer to live in areas with plenty of water for at least part of the year, although they will stray on to grasslands and into deserts in search of food. They live alone, taking refuge in secluded spots during the day and stalking prey at night. Despite being expert climbers, they hunt on the ground and drag their kills to hideaways before devouring them.

Female jaguars defend smaller territories than males, and a male's territory may overlap those of two or three females. The cats advertise

their presence by scenting landmarks with urine or faeces and by scraping marks on tree trunks and rocks. When a female is ready to breed, she will leave her home range and be courted by outside males. Litters usually stay with their mother for about two years.

Distribution: South-western United States, Mexico, Central and South America to northern Argentina.
Habitat: Forests and swamps.
Food: Capybaras, peccaries, caimans and tapirs.
Size: 1.1–1.9m (3.5–6.25ft); 36–158kg (80–350lb).
Maturity: 3 years.
Breeding: Litters of 1–4 cubs born every 2 or 3 years.
Life span: 22 years.
Status: Lower risk.

Pampas cat (*Felis colocolo*): 57–70 cm (23–28in); 2–3kg (4.5–6.5lb)
The pampas cat is a small cat with a long coat of yellowish or brown striped fur. This rare species lives in the South American treeless grasslands known as pampas, most of which are in Argentina. Pampas cats hide during the day in the tall grasses of the pampas, but may hide in one of the rare trees in the region. At night the cat feeds on guinea pigs and ground birds.

Bobcat (*Felis rufus*): 65–100cm (25–40in); 4.1–15.3kg (9–33lb)
Bobcats live across the southern half of North America. They are found from semi-deserts to mountain forests, making their dens in tree hollows and rock crevices. Bobcats hunt at night, patiently stalking prey over ground before attacking with a deadly pounce. Territories are marked with the scent from faeces, urine and a gland above the anus. Mating takes place in late winter or early spring, with litters of up to six cubs staying with their mother until the next winter.

Jaguarundi (*Felis yagouaroundi*): 55–77cm (22–30in); 4.5–9kg (10–20lb)
The jaguarundi is a small forest cat with a grey-brown coat found in the lowlands of Central and South America. Unlike most cats, jaguarundis are crepuscular – they hunt in the twilight of dawn and dusk when prey are disorientated by the changing light levels. Their diet consists mainly of birds and small rodents. Reports suggest that some jaguarundis live in pairs in parts of their range, but they otherwise appear to be solitary animals.

Ocelot

Felis pardalis

Ocelots are medium-sized small cats found across most of the American tropics. These agile hunters are most common in the dense jungles of the Amazon Basin, but are also found high on the slopes of the Andes and in the dry shrublands of northern Mexico.

An ocelot's typical day is spent sleeping in the cool of a shady thicket or on a leafy branch, but at night the cat comes out to hunt. Ocelots eat a wide range of animals, including rodents, snakes and even young deer and peccaries if the opportunity arises.

Ocelots are largely solitary animals, although males will maintain social links with numbers of females in their local areas. They communicate with quiet mews, which become loud yowls during courtship. In the heart of the tropics, ocelots breed all year round, while at the northern and southern extremes of their range, they tend to mate during the late summer and fall.

Ocelots used to be hunted for their fur. The cats are now protected from hunting but are still threatened by deforestation.

Distribution: Mexico to northern Argentina.
Habitat: Tropical forest.
Food: Rodents, rabbits, birds, snakes and fish.
Size: 55–100cm (22–40in); 11.5–16kg (25–35lb).
Maturity: 18 months.
Breeding: Litters of 2–4 born once a year.
Life span: 15 years.
Status: Lower risk.

DOGS

Domestic dogs belong to the Canidae *family, which they share with similar animals, such as wolves (from which they are descended), foxes and jackals. Most types of wild canid live in large family groups called packs. Dog societies are very complex because the animals must cooperate to survive. They hunt together and take it in turns to care for the young.*

Maned wolf

Chrysocyon brachyurus

The maned wolf lives in areas of swamp and open grassland in central South America, east of the Andes. Its name comes from a dark swathe of hair on its nape and along its spine. The hairs in this mane stand erect when the animal is threatened.

Maned wolves form monogamous pairs throughout their lives. Males and females share territory and have dens hidden inside thick vegetation. Most of the time they stay out of each other's way, hunting alone at night. The pair only spend time together during the breeding season – at the end of winter. Both parents help to raise the litter, regurgitating food at the den for the young to feed on.

Unlike other wolves, which run down their prey, maned wolves stalk their victims more like foxes. Despite having very long legs, maned wolves are not great runners. Instead, their height allows them to peer over tall grasses looking for prey.

Maned wolves have fox-like coloration, with a reddish-brown coat of longish fur. These canids are omnivorous, supplementing their diet with fruit.

Distribution: Central and eastern Brazil, eastern Bolivia, Paraguay, northern Argentina and Uruguay.
Habitat: Grassland.
Food: Rodents, other small mammals, birds, reptiles, insects, fruit and other vegetable matter.
Size: 1.2–1.3m (4–4.5); 20–23kg (44–50lb).
Maturity: 1 year.
Breeding: Monogamous pairs produce litters of 2–4 cubs.
Life span: 10 years.
Status: Lower risk.

Grey wolf

Canis lupus

Grey wolves howl to communicate with pack members over long distances. Each individual can be identified by its howl.

All domestic dogs are descended from grey wolves, which began living alongside humans many thousands of years ago. Grey wolves are the largest dogs in the wild, and they live in packs of about ten individuals. A pack has a strict hierarchy, with a male and female "alpha pair" in charge. The alpha dogs bond for life and are the only members of the pack to breed. The rest of the pack is made up of the alpha pair's offspring.

In summer, pack members often hunt alone for small animals such as beavers or hares, while in winter, the pack hunts together for much larger animals. Grey wolves are strong runners and can travel 200km (125 miles) in one night. They generally detect prey by smell and chase them down, taking turns to take a bite at the faces and flanks of their victims until they collapse from exhaustion.

Distribution: Canada and some locations in the United States and Europe, and across most of Asia.
Habitat: Tundra, pine forest, desert and grassland.
Food: Moose, elk, musk ox and reindeer.
Size: 1–1.6m (3.25–5.25ft); 30–80kg (66–175lb).
Maturity: 22 months.
Breeding: Once per year.
Life span: 16 years.
Status: Vulnerable.

Kit fox

Vulpes macrotis

Distribution: Western United States.
Habitat: Desert and scrub.
Food: Rodents, pikas, insects and fruit.
Size: 38–50cm (15–20in); 1.9–2.2kg (4–5lb).
Maturity: 1 year.
Breeding: Litters of 4–5 cubs.
Life span: 15 years.
Status: Vulnerable.

Kit foxes live in the dry desert and scrub areas of the high plateaux and valleys beside the Rocky Mountains in the United States. They generally live in breeding pairs, but social bonds are quite loose and pairs often split. The female does not leave her den – in a disused burrow – while she is suckling her litter of four or five cubs. During this time she relies on the male for food, which is generally small rodents and rabbits, insects and fruit.

After three or four months, the young are strong enough to travel with their parents to other dens in their territory. A kit fox family's territory overlaps widely with those of other groups in the area. The size of the territory depends on the climate. Desert territories have to be large to supply enough food for the family. The kit fox is very similar in appearance and behaviour to the swift fox (*Vulpes velox*) which lives on the great plains farther east. It is possible that hybridization takes place where the ranges of these two dogs overlap.

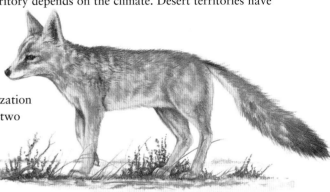

The kit fox's large ears are lined with blood vessels that radiate heat to cool the animal down in hot desert climes.

Hoary fox (*Dusicyon vetulus*): 59–64cm (23–25in); 4kg (9lb)
Hoary foxes live in south-central Brazil. They have short coats of yellow and black hairs. They inhabit sparsely wooded savannahs, taking shelter in burrows deserted by other animals, such as armadillos. The foxes do hunt down small mammals and birds, but a lot of their diet is made up of insects, such as grasshoppers. For this reason, their grinding molar teeth are wider than those of other dogs.

Crab-eating fox (*Dusicyon thous*): 60–70cm (24–28in); 6–7kg (13–15lb)
Crab-eating foxes live in the woodlands and grasslands of western South America that exist in the highlands around the Amazon Basin. They feed on both coastal and freshwater crabs as well as insects, fruit and carrion. Crab-eating foxes are nocturnal and they locate crabs in the dark by listening for rustling noises among thick vegetation. This species lives in pairs and may breed at any time of the year.

Coyote (*Canis latrans*): 75–100cm (30–40in); 8–20kg (18–44lb)
Coyotes live across the whole of North America, from the deserts of Mexico to the barren tundra near the Arctic. These largely solitary dogs look a little like small grey wolves. Coyotes are active in the dark, and especially at dawn and dusk, but they do sometimes hunt during the day. Coyotes are the fastest runners in North America, reaching speeds of 64kph (40mph). This helps them catch swift jackrabbits – their favourite food.

Bush dog

Speothos venaticus

Bush dogs are unusual members of the dog family, looking more like weasels or mongooses than other dogs. They live in wetlands and flooded forests in highly social packs of about ten dogs. Pack members hunt together, chasing ground birds and rodents. Like other pack dogs, their victims are often much bigger than an individual dog, for example capybaras, agoutis and rheas. They are thought to be expert swimmers, diving into water as they chase prey.

Bush dogs are diurnal – active during the day – and keep together by making high-pitched squeaks as they scamper through the dense forest. As night falls, the pack retires to a den in a hollow tree trunk or abandoned burrow. Little is known about the social system within the packs, but it is likely that there is a system of ranking.

Litters of two or three young are produced in the rainy season. Females only become ready to breed once they come into contact with males.

Distribution: Northern South America, east of the Andes.
Habitat: Forests and swampy grasslands.
Food: Ground birds and rodents.
Size: 57–75cm (22–30in); 5–7kg (11–15lb).
Maturity: 1 year.
Breeding: Litter of 2–3 cubs born in rainy season.
Life span: 10 years.
Status: Vulnerable.

A whole pack of bush dogs squeezes into one den to spend the night.

BEARS

The world's largest land carnivore, the Kodiak bear, lives in North America. It is a huge, hairy animal that can grow to 3m (10ft) tall. Despite their immense size and strength, bears are generally not the vicious predators many people think they are. Most eat more plant food than meat, and they are usually shy beasts, preferring to stay away from humans.

Polar bear
Ursus maritimus

Distribution: Arctic Ocean to southern limits of floating ice, and Hudson Bay.
Habitat: Ice fields.
Food: Seals, reindeer, fish, seabirds and berries.
Size: 2–2.5m (6.5–8.25ft); 150–500kg (330–1100lb).
Maturity: 6 years.
Breeding: 1–4 cubs born every 2–4 years.
Life span: 30 years.
Status: Vulnerable.

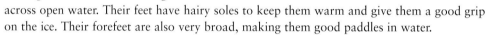

Polar bears have proportionally longer necks than other types of bear so that they can lunge after seals and other aquatic prey.

Polar bears are semi-aquatic animals. They live on the fringes of the vast ice fields that surround the North Pole, where they feed on seals and other marine animals. The bears may cover large distances in search of food, sometimes coming far inland or swimming for miles across open water. Their feet have hairy soles to keep them warm and give them a good grip on the ice. Their forefeet are also very broad, making them good paddles in water.

The bears' snow-white coats help them blend into their surroundings and stay hidden from their prey. A polar bear's staple food is the ringed seal. The bears either wait beside holes in the ice for seals to surface, or sneak up on them across the ice. Bears sometimes dig down into seal dens beneath the surface snow and eat new-born pups.

Polar bears put on a lot of weight in summer because they have less opportunity to feed in winter. They often take shelter from extreme weather in underground dens. Pregnant females sleep inside large dens for long periods during the winter months, before giving birth to their pups in spring. The young stay with their mothers for two years.

Brown bear
Ursus arctos

Brown bears live in many parts of the northern hemisphere, and although they belong to a single species, they look rather different from place to place. For example, the brown bears in Europe and Asia are smaller and darker than their American cousins. In North America, there are two races of brown bear: Kodiaks and grizzlies.

Brown bears make their homes in cold places, such as northern forests, mountains and barren tundra. They feed on a range of fruits, plants and small animals. Only grizzlies regularly attack larger animals, such as deer and even smaller black bears.

Distribution: North America, Siberia, Europe and Caucasus Mountains.
Habitat: Tundra, alpine meadows and forests.
Food: Salmon, grasses, roots, mosses, bulbs, insects, fungi, rodents, deer, mountain sheep and black bears.
Size: 1.7–3m (5.5–10ft); 100–700kg (220–1,540lb).
Maturity: 6 years.
Breeding: 1–4 cubs born every 3–4 years.
Life span: 25–30 years.
Status: Endangered in some places.

Brown bears are generally solitary animals, although they may gather in groups around large food supplies, such as schools of salmon beneath waterfalls. As winter approaches, the bears dig themselves dens for semi-hibernation. Although they sleep during most of the winter, they often come out of the den for short periods between sleeps. Mating takes place in early summer. The female gives birth in spring, and her cubs stay with her for at least two years.

Brown bears have humps between their powerful shoulders, and longer claws than most other bears.

American black bear

Ursus americanus

Distribution: Alaska, Canada and patchily throughout parts of the United States from New England to Tennessee, Florida, Mississippi and western states into northern Mexico.
Habitat: Forests.
Food: Fruits, nuts, grass, roots, insects, fish, rodents and carrion.
Size: 1.3–1.8m (4.25–6ft); 100–270kg (220–594lb).
Maturity: 6 years.
Breeding: 1–5 cubs born every 2 years.
Life span: 25 years.
Status: Lower risk.

American black bears are the smallest bears in North America. They live in the conifer forests of Canada and a few wilderness areas as far south as Mexico. They share these forests with grizzly bears and are sometimes eaten by them. Their main defence against this is to climb trees out of the reach of the less agile grizzly.

Black bears are most active at night. Three-quarters of what they eat is plant matter, with small animals, such as fish and rodents, making up the rest. Like other bears, black bears semi-hibernate through the winter in dens under fallen trees or in burrows. Although they sleep heavily, they often wake through the winter, going on excursions during breaks in the severe winter weather.

Although black bears generally forage for food alone, they will congregate around a large source of food. In general they stay away from each other, especially unknown bears. In the middle of summer, males and females come together for short periods. The male leaves soon after mating and cubs are born at the end of winter, while the mother is still in her winter den. The young stay with their mother until at least two years old, usually when they are driven away by the aggression of males courting their mother.

American black bears vary in coloration from black to dark or reddish-brown and pale tan. They differ from grizzlies in several respects; for example, they have shorter fur and they lack a shoulder hump. They also have shorter legs and claws, which helps them to climb trees with far more adeptness than grizzlies, and so escape these predators. They vary in size according to the quality of food available.

Spectacled bear

Tremarctos ornatus

Distribution: Northern Andes Mountains, including Colombia, Ecuador, Peru, Bolivia and into Chile.
Habitat: Tropical mountain forest and alpine grassland.
Food: Fruit, epiphytes, bamboo hearts, corn, rodents and insects.
Size: 1.2–1.8m (4–6ft); 60–175kg (132–385lb).
Maturity: Not known.
Breeding: 2 cubs born every 2–3 years.
Life span: 25 years.
Status: Vulnerable.

The spectacled bear is the only species of bear in South America. It lives mainly in the lush, high-altitude forests clothing the slopes of the Andes Mountains from Colombia southward as far as northern Chile.

Spectacled bears are active at night, especially during the twilight hours. During the day they shelter in caves, under tree roots or on tree trunks. They are expert climbers and spend a great deal of time foraging in trees. Once up trees, the bears often build feeding platforms from broken branches. They use these platforms to reach more food.

The spectacled bear's food is mainly fruit, and it will travel through the forest collecting ripe fruits. During periods when ripe fruit is unavailable, the bears eat epiphytes – plants that grow on other plants – called bromeliads, feasting on the soft edible hearts of the plants.

Being a tropical species, breeding occurs all year round. Pairs stay together for a few weeks after mating, but the cubs are born seven months later. The cubs stay with their mother for at least two years before being chased away by adult males seeking to mate with their mother.

Spectacled bears are so named because of the large white circles or semicircles of whitish fur around their eyes.

SMALL CARNIVORES

Most small carnivores belong to the Mustelidae *family. The mustelids are a diverse group, including otters, martens and badgers, which are adapted to aquatic, arboreal and subterranean lifestyles respectively. The world's largest and most successful mustelids live in the Americas, where they are found from the icy north to the humid tropics.*

Striped skunk

Mephitis mephitis

The striped skunk is well known for the foul-smelling spray it produces to ward off attackers. This spray comes out of two tiny apertures inside the anus. The discharge, known as musk, is squirted in spray form or as a directed arc of droplets.

The skunk will only spray when it has exhausted all other defensive tactics. These include arching its back, holding its tail erect and stamping its feet. If these fail, the skunk will twist its body into a U-shape – so its head and tail are facing the attacker – and release its musk. The musk, which can be smelled by humans over a mile away, causes discomfort to the eyes of an enemy.

Striped skunks are most active at night, foraging for food under the cover of thick vegetation. They spend the day in sheltered places, such as disused burrows. During the winter, skunks hibernate in their dens, staying underground for between two to three months. Mating takes place in springtime. Litters of up to ten young are born in summer.

The striped skunk is characterized by the broad white stripes that extend from the top of its head to the tip of its tail.

Distribution: North America.
Habitat: Woods, grasslands and deserts.
Food: Rodents, other small vertebrates, insects, fruit, grains and leaves.
Size: 28–30cm (11–12in); 0.7–2.5kg (1.5–5.5lb).
Maturity: 1 year.
Breeding: 1–10 young born every summer.
Life span: 6 years.
Status: Common.

American mink

Mustela vison

American mink are small carnivores that live close to water, where they feed on small aquatic animals. They originally came from North America, but were brought to Europe and Asia to be farmed for their fine fur. They have since escaped into the wild and are now a common pest. They are also competition for the similar, but very rare, European mink.

Mink prefer to live in areas with plenty of cover. Their river-bed dens are generally deserted burrows made by other river mammals, but mink will dig their own when necessary.

Mink are active at night and dive into water to snatch their prey. They live alone and will defend their own stretches of riverbank against intruders. A litter of up to five young is born two months after mating inside dry, underground nests lined with fur, feathers and leaves. In autumn the young begin to fend for themselves.

Mink are known for their luxurious fine fur, which is used for clothing. Several domestic varieties of mink have been bred, each with different- colored fur.

Distribution: North America. Introduced to northern Europe.
Habitat: Swamps and near streams and lakes.
Food: Small mammals, fish, frogs and crayfish.
Size: 33–43cm (13–17in); 0.7–2.3kg (1.5–5lb).
Maturity: 1–1.5 years.
Breeding: 5 young born in late spring.
Life span: 10 years.
Status: Common.

Wolverine

Gulo gulo

Wolverines are giant relatives of the weasels. As well as the conifer forests of North America, they are found in northern Europe and Siberia, where they are known as gluttons due to their catholic feeding habits.

Wolverines are generally nocturnal, but will forage by day if they need to. Their diet varies throughout the year. In summer, they feed on small animals, such as rodents and ground-living birds, and readily feast on summer fruits. In winter, when most other carnivores are hibernating, wolverines may tackle bigger prey, such as deer. Their wide feet act as snowshoes, allowing them to walk over deep snow, in which hapless deer are defenceless and unable to make their escape.

Wolverines mate in early summer and young are born in underground dens the following spring. They leave their mothers in the autumn.

Wolverines have large heads and heavily built bodies with dense coats of hairs of different lengths to prevent winter snow and ice from getting too close to the skin, causing heat loss.

Distribution: Canada, northern United States, Scandinavia and Siberia.
Habitat: Tundra and conifer forest.
Food: Carrion, eggs, rodents, berries, deer and sheep.
Size: 65–105cm (25–41in); 10–32kg (22–70lb).
Maturity: 2–3 years.
Breeding: Litter of 2–4 born in early spring every 2 years.
Life span: 10 years.
Status: Vulnerable.

Spotted skunk (*Spilogale putorius*): 12–35cm (5–14in); 0.2–1kg (0.5–2.2lb)
This small skunk lives throughout most of North America. It has a unique pattern of white stripes and spots on its rump. Like the larger American skunks, spotted skunks spray foul-smelling musk. They spend the day under cover, sometimes hiding in trees. They do not hibernate, and eat a mixed diet including plants, insects and rodents. Spotted skunks mate in autumn and give birth the following spring to four or five young.

Black-footed ferret (*Mustela nigripes*): 38–50cm (15–20in); 0.7–1kg (1.5–2.2lb)
This extremely rare carnivore lives in just a few dry prairies in Wyoming, USA. Black-footed ferrets live in the burrows – called towns – of prairie dogs. The ferrets rest during the day and then travel around the town at night, preying on the resident prairie dogs. Mating takes place in spring and the young are born six weeks later.

Greater grisón (*Galictis vittata*): 47–55cm (18–22in); 1.4–3.3kg (3–7lb)
The greater grisón is one of only a few South American mustelids. It ranges from southern Mexico to central Peru and south-eastern Brazil, making its home in both forest and open country. Grisóns have grey backs with a thin white stripe extending down the middle. They make their dens in burrows abandoned by viscachas and other large rodents, and appear to be active during both the day and night. Grisóns feed on small mammals, birds, eggs, lizards and frogs. Breeding takes place throughout the year.

American badger

Taxidea taxus

American badgers are tough animals that live in the open country in the Great Plains region of North America. They are expert burrowers and use this skill to dig out their preferred foods – rodents, such as prairie dogs and ground squirrels. They rest in their burrows during the day and come out to feed at night. During the coldest weeks of the year, American badgers do not hibernate, but sleep underground for several days at a time.

They may bury some of their food for later or even dig holes big enough for them and their prey to fit into. American badgers and coyotes are known to hunt together in teams. The coyotes sniff out the buried prey and the badgers dig them out. Both parties then share the food.

Mating occurs in summer and early autumn, and births take place in the following spring. The young leave home after two months.

American badgers have a single white stripe running from the nose along the back. Northern-race badgers have the stripe to their shoulders, while on those of the south race it runs all the way along the back.

Distribution: Central and southern North America.
Habitat: Dry, open country.
Food: Rodents, birds, reptiles, scorpions and insects.
Size: 40–70cm (16–28in); 4–12kg (9–26.5lb).
Maturity: Females 4 months; males 1.3 years.
Breeding: 1–5 young born in spring.
Life span: 14 years.
Status: Lower risk.

Sea otter

Enhydra lutris

Sea otters live in the cold coastal waters around the northern Pacific Rim. They do not need to come on to land to survive, but often do. Unlike other marine mammals, sea otters do not have thick blubber under their skins for insulation. Instead, they rely on a layer of air trapped by their soft fur to insulate them against the cold. Pollution, such as oil in the water, can reduce the fur's ability to trap air, and otters may die of hypothermia as a result.

The otters spend a minute or two at a time underwater, collecting food such as shellfish and urchins. They then float on their backs to feed. They smash the hard shells against stones to get at the soft meat inside, using their chests as tables.

Sea otters are active during the day. At night they wrap themselves in kelp before going to sleep to prevent themselves from floating away. They sometimes put their forepaws over their eyes while sleeping.

Sea otters live alone and only tolerate each other when mating. Breeding takes places all year round. Pups are carried on the mother's chest until they are two months old, when they begin to feed themselves.

Distribution: Northern Pacific coasts from California to Japan.
Habitat: 20m (60ft) deep water.
Food: Fish and shellfish, such as sea urchins, abalones, crabs and molluscs.
Size: 1–1.2m (3.25–4ft); 15–45kg (33–99lb).
Maturity: Females 4 years; males 6 years.
Breeding: Single pup every 1–3 years.
Life span: 20 years.
Status: Threatened.

Sea otter fur comprises 100,000 hairs per 1sq cm (0.15sq in), making it the densest of any mammal. This keeps the animal warm in the cold ocean. The hind feet are webbed and flipper-shaped.

Giant otter

Pteronura brasiliensis

The giant otter is the largest mustelid in the world, although it is not as heavy as the sea otter. This semi-aquatic mammal inhabits the tropical river basins of South America. It lives in groups of about six, each communicating with chirping sounds. Generally, the group comprises an adult pair and their offspring of various litters. Each group controls its own stretch of stream, preferring those areas with plenty of cover.

The giant otter swims at high speed by waving its tail and body up and down, using its webbed feet to steer. On land it is far less agile, and is often seen sitting grooming itself. Giant otters are diurnal – only active during the day. They catch prey in their mouths and hold it in their forepaws to eat it on the shore. During the dry season, the otter groups are restricted to small areas of water, but when the rains come to flood the forest, the otters can roam over larger areas. Little is known about the mating habits of giant otters, other than that the young stay with their parents for a few years before reaching adulthood.

Distribution: Central America and South America from Venezuela to Argentina.
Habitat: Slow-moving rivers and creeks in forests and swamps.
Food: Fish, fish eggs, crabs, birds and small mammals.
Size: 0.8–1.4m (2.5–4.75ft); 22–34kg (48–75lb).
Maturity: Not known.
Breeding: 1–3 young produced every year.
Life span: 15 years.
Status: Vulnerable.

The giant otter's fur has a velvety appearance, more like the pelt of a seal than an otter. Its feet are large and have thick webbing, and the tail is flattened into a flipper-like shape.

North American river otter

Lutra canadensis

Distribution: North America.
Habitat: Rivers and lakes.
Food: Amphibians, fish, crayfish and aquatic insects.
Size: 60–110cm (24–43in); 3–14kg (6.5–30lb).
Maturity: 2–3 years.
Breeding: 1–5 young born every year.
Life span: 20 years.
Status: Lower risk.

North American river otters rarely stray far from the banks of shallow rivers. They live alone or in pairs, but often play with other individuals in the area. This play strengthens social ties. Each of the otters has an individual scent which it uses to mark its territory. River otters communicate with each other through sounds such as whistles, growls, chuckles and screams.

North American river otters are known for their boundless energy, and they must eat frequently. They catch fish in their mouths and detect other prey by feeling with their whiskers along the bottoms of streams. Unlike many other otters which chew their food, the river otter's prey is gulped down immediately.

Mating takes place in March and April. The young are born almost a year later. The females give birth in dens close to the water's edge. They drive the males away soon after the birth of their young, but the dog otters return later to help raise the offspring. The young depart at the age of one year.

River otters have streamlined bodies with dark fur, thick tails and short legs with webbed feet.

Patagonian weasel (*Lyncodon patagonicus*): 30–35cm (12–14in); weight not known
This mustelid has grey-brown fur, with broad white stripes running from the head to the shoulders. The weasel's body is similar to that of a grisón, except that the tail is shorter. The Patagonian weasel lives in the pampas – grasslands of Argentina and southern Brazil. Little is known about how this species behaves, but its large cutting teeth suggest that it is highly predaceous.

Tayra (*Eira barbata*): 56–68cm (22–27in); 4–5kg (9–11lb)
This unusual species is only found in central Mexico and on the island of Trinidad. Its coat is short and varies from grey to black. Typical among mustelids, the tayra has a slender body, short limbs and a long tail. This carnivore is found in dense forest, living in small groups on the ground, preying on rodents and small deer. When it is being chased by predators, it has been known to run up trees and leap through the branches before coming back down at another location.

American pine marten (*Martes americana*): 32–45cm (13–18in); 0.3–1.3kg (0.5–3lb)
The American pine marten lives in the cold northern forests of Canada and high-altitude areas of the United States. It has a dark coat and a long slender body. Pine martens spend the day in nooks and crannies in the forest, and move through the trees and along the ground in search of food during the night. Their diet includes small mammals, carrion, fruit and insects.

Fisher

Martes pennanti

The fisher, or pekan, lives in the thick forests of North America. Despite its name, it feeds on small land animals, such as mice and porcupines. Fishers have no permanent dens, but take shelter in hollow trees, holes in the ground and even abandoned beaver lodges.

They are active during the day and night, and despite being expert climbers, spend most of their foraging time on the ground. When they come across suitable prey animals, they rush forward and kill them with bites to the back of the neck. Larger animals are killed with repeated bites to the face.

Males seek out mates during the spring breeding season and litters are born about ten months later. As with many mustelids, the fertilized eggs do not begin to grow immediately inside the females. Their development is delayed for several months so that they are born at the right time of year. Unusually, births always take place in trees.

Distribution: Canada and northern United States.
Habitat: Conifer forest.
Food: Birds, rodents, carrion.
Size: 49–63cm (19–25in); 1.3–3.2kg (3–7lb).
Maturity: 1–2 years.
Breeding: 3 young born every spring.
Life span: 10 years.
Status: Lower risk.

Fishers have dark fur that is coarser than that of most mustelids. Nevertheless, they are still hunted by humans for their fur.

RACCOONS AND RELATIVES

Raccoons and their relatives belong to a family of mammals called the Procyonidae. *Procyonids are small opportunistic feeders and scavenging animals. Many live in trees, but the most successful – the raccoons – live mainly on the ground. Most procyonids live in the Americas, where they range from the cold northern forests of Canada to the humid, tropical swamps of the Amazon.*

Common raccoon

Procyon lotor

Raccoons live in woodland areas and rarely stray far from water. They are more active at night than during the day. Periods of rest are spent in dens in tree hollows or other sheltered places. When on the move, raccoons will readily swim across streams and rivers and will climb into trees in search of food. They use their touch-sensitive hands to grab prey and then break it into mouth-sized pieces.

Raccoons do not hibernate in warmer parts of their range, although in cooler northern parts they may do so. In fact, they only semi-hibernate, popping out every now and then to feed during breaks in the severest weather.

Males are largely solitary but will tolerate the presence of females living in or near their territories. Mating takes place in spring, and young are born a couple of months later. The young stay with their mothers until the following spring.

The common raccoon is well known for its black "bandit" mask across the eyes and its tail ringed with black hoops. The animal's footprints look similar to those of a human infant.

Distribution: Southern Canada throughout the United States to Central America.
Habitat: Forests and brushland.
Food: Crayfish, frogs, fish, nuts, seeds, acorns and berries.
Size: 41–60cm (16–24in); 2–12kg (4.5–26.5lb).
Maturity: 1 year.
Breeding: 3 or 4 young born in summer.
Life span: 5 years.
Status: Common.

Olingo

Bassaricyon gabbii

Distribution: Central America to northern South America as far as Brazil.
Habitat: Tropical forest.
Food: Fruit, insects and small mammals.
Size: 35–48cm (14–19in); 0.9–1.5kg (2–3.5lb).
Maturity: 21 months.
Breeding: Single offspring.
Life span: 5 years.
Status: Lower risk.

Olingos live in the trees of tropical forests. They are active at night and spend the day in nests of leaves high up inside hollow trees. Equipped with long claws, olingos are expert climbers, and they rarely descend to the ground. They can also jump long distances through the treetops, using their long tails to keep them balanced.

An olingo's diet comprises mainly fruit, although it will seek out insects and small vertebrates, such as lizards, on occasion. Olingos live alone, although they are often found living alongside kinkajous – procyonids that are close relatives – as well as opossums and night monkeys.

They mark objects in their territories with urine, although it is not known whether this is to ward off intruders or to help them navigate in the darkness. Mating takes place all year round. Gestation lasts about ten weeks and generally results in a single offspring.

Olingos have thick, pinkish fur. These procyonids have long bodies with short limbs and flattened tails.

Ring-tailed coati

Nasua nasua

Distribution: Northern South America as far as Argentina.
Habitat: Woodland.
Food: Fruit, insects, rodents.
Size: 41–67cm (16–26in); 3–6kg (6.5–13lb).
Maturity: 2 years.
Breeding: 2–7 young born in rainy season.
Life span: 10 years.
Status: Common.

Coatis have long muzzles compared to raccoons and other procyonids. They use these to root out food from rocky crevices and from knots in trees. Coatis forage both on the ground and in trees. On the ground they hold their long tails erect, with the tips curled. In trees, coatis' tails are prehensile enough to function as a fifth limb. The tips curl around branches to provide support in more precarious locations.

Coatis are most active during the day. When there is plenty of fruit on the trees, coatis will eat little else. However, during seasons when fruit is less abundant, coatis comes down to the forest floor to forage for insects and rodents.

Coatis tend to congregate in bands of up to 20 females and young. Adult males live alone and are only allowed into bands during the breeding season, which is the time when there is plenty of fruit available. When fruit is not as easy to find, male coatis may try to eat smaller members of their band, and consequently are expelled by the adult females.

Like all coatis, this species has a long and pointed muzzle with an articulated tip. Ring-tailed coatis have long, coarse fur, and tails banded with white stripes.

Cacomistle (*Bassariscus astutus*): 30–42cm (12–16in); 0.8–1.3kg (1.75–3lb)
Also called ringtails, cacomistles have small cat-like bodies and bushy black and white striped tails. They live in rocky areas from Oregon to Kansas, and southward to southern Mexico. They seldom stray far from water and are most active at night, foraging mainly for insects, rodents and fruit. Being agile climbers, they move up and down cliffs with ease. Like those of many other climbing animals, a cacomistle's hind feet can twist around 180 degrees. This allows the animals to climb down trees and rocks head-first, with the claws on the hind feet clinging to the surface behind them.

Mountain coati (*Nasuella olivacea*): 35–45cm (14–18in); 2kg (4.5lb)
Mountain coatis closely resemble other coatis, but tend to be smaller and have shorter tails. They are very rare and live in the tropical forests on the slopes of the Andes Mountains in South America. They feed on insects, fruit and small vertebrates, which they find among the trees and on the ground.

Crab-eating raccoon (*Procyon cancrivorus*): 45–90cm (18–35in); 2–12kg (4.5–26lb)
Crab-eating raccoons, or mapaches, live in swamps or by streams across most of South America, east of the Andes Mountains and north of Patagonia. This species has much shorter hair and a more slender body than most of its raccoon cousins. They search the water for food at night, detecting prey – shellfish, fish and worms – with their touch-sensitive paws.

Kinkajou

Potos flavus

Kinkajous are almost entirely arboreal (tree-living). Thanks to their long claws and prehensile tails, they are very agile climbers. Kinkajous are nocturnal and spend the day in dens inside hollow trees. On the hottest days they emerge from their stifling dens to cool off in the open on branches.

At night, kinkajous race around the trees in search of fruit. After searching through one tree, they will cautiously move to the next before beginning to forage again. They use their long tongues to reach the soft flesh and juices inside the fruit.

Kinkajous tend to return to the same roosting trees each dawn. They travel alone or in breeding pairs. However, groups of kinkajous may form in trees that are heavy with fruit. Kinkajous leave their scent on branches, probably as a signal to potential mates. They also give shrill calls to communicate with partners. Mating takes place all year round, and single offspring are born after four months.

Distribution: Mexico to central Brazil.
Habitat: Forests.
Food: Fruit, insects and small vertebrates.
Size: 40–76cm (16–30in); 1.4–4.6kg (3–10lb).
Maturity: 1.5–2.5 years.
Breeding: Single offspring.
Life span: 15 years.
Status: Endangered.

Kinkajous have soft and woolly fur, with rounded heads and stockier bodies than most of their relatives. They are sometimes mistaken for African primates called pottos.

SQUIRRELS AND RELATIVES

Squirrels and their close relatives, such as beavers, marmots and gophers, are all types of rodent, so they are related to rats and mice. The squirrel family numbers over 350 species in two main groups: the ground squirrels that dwell on or under the ground, and the arboreal species – tree and flying squirrels. American squirrels include three ground-dwellers: the woodchuck, beaver and prairie dog.

Grey squirrel

Sciurus carolinensis

Distribution: Eastern North America. Introduced to parts of Europe.
Habitat: Woodlands.
Food: Nuts, flowers and buds.
Size: 38–52cm (15–21in); 0.3–0.7kg (0.5–1.5lb).
Maturity: 10 months.
Breeding: 2 litters born each year with 2–4 young per litter.
Life span: 12 years.
Status: Common.

Grey squirrels are native to the open woodlands of eastern North America. They have also been introduced into parts of Europe, where they have out-competed the smaller red squirrels for food and breeding sites.

Grey squirrels feed primarily on the nuts and buds of many woodland trees. In summer, when they are most active just after dawn and before dusk, grey squirrels also eat insects. In winter, when most animals of their size are hibernating, grey squirrels spend their days eating stores of food which they buried throughout the previous summer. Grey squirrels may make dens in hollow trees, but are more likely to make nests, or dreys, from twigs and leaves in the boughs of trees.

There are two breeding seasons each year: one beginning in midwinter, the other in midsummer. Males begin to chase females through the trees a few days before they are receptive to mating. When females are ready, their vulvas become pink and engorged. Litters of three are born six weeks later.

Grey squirrels have, as their name suggests, greyish fur, although many individuals have reddish patches. Their tails, which have many white hairs, are bushier than those of most other squirrels.

Woodchuck

Marmota monax

Woodchucks are the largest squirrels in North America. Their powerful legs have curved claws, which are used for digging.

Woodchucks are also called groundhogs or whistlepigs – the latter because of the shrill alarm call they make when threatened. Unlike most other squirrels, they eat the green parts of plants rather than the seeds and buds. They also eat bark and small twigs. Their natural habitat is the edge of forests or other open areas where there is plenty of cover. With the growth of agriculture, woodchucks have increased in number, making use of hedges beside open fields. They live alone, unlike most other ground squirrels.

In winter, woodchucks hibernate by using subcutaneous fat reserves they have put on over summer. Their winter sleep is much deeper than most squirrels. Mating takes place soon after hibernation ends. Female woodchucks have a single litter every year and males mate with more than one female. Young woodchucks are thrown out of their mother's burrow after a few months.

Distribution: Southern Canada southward through eastern North America.
Habitat: Woodland or open areas that have plenty of ground cover.
Food: Plant leaves and stems.
Size: 45–65cm (18–26in); 2–5kg (4.5–11lb).
Maturity: 2 years.
Breeding: 3–5 young born in May.
Life span: 6 years.
Status: Common.

Northern pocket gopher

Thomomys talpoides

Pocket gophers have robust, tubular bodies with short legs. Their forefeet have long claws and their tails are naked at the tip. Male gophers are much larger than females.

Pocket gophers spend a great deal of their time burrowing. They feed on the underground parts of plants, such as roots, tubers and bulbs. The gophers access their food by digging temporary feeding tunnels out and up from deeper and more permanent galleries, located 1–3m (3.25–10ft) underground. Gophers keep their burrow entrances blocked with earth most of the time, and rarely appear above ground during the day. At night they may move around on the surface.

Gophers carry food in pouches inside their cheeks to storage or feeding sites in their burrow systems. They do not drink water, and so get all of their liquid from plant juices.

Only during the mating season will a male be allowed into a female's burrow. Litters are born just 18 days after mating, which generally takes place in summer.

Distribution: Western North America from Canada to Mexico.
Habitat: Burrows under desert, prairie and forest.
Food: Roots, bulbs and leaves.
Size: 11–30cm (4.5–12in); 50–500g (0.1–1.1lb).
Maturity: 1 year.
Breeding: 1–10 young born in summer.
Life span: 2 years.
Status: Common.

Flying squirrel (*Glaucomys volans*): 21–70cm (8–28in); 50–180g (0.1–0.4lb)
Flying squirrels range over eastern and central North America, from Quebec to Honduras. Their bodies resemble those of other squirrels, except that they have loose folds of skin along the sides of their bodies and attached to their elongated arms and legs. This membrane is used for gliding from tree to tree.

Prairie dog (*Cynomys ludovicianus*): 28–33cm (11–13in); 0.7–1.4kg (1.5–3lb)
This species of prairie dog, and others like it, live in open grasslands from western Canada to Mexico. They are not true dogs, but ground squirrels that make a barking call. They live in colonies inhabiting communal burrows called towns. In frontier times, one huge town in west Texas was estimated to contain 400 million prairie dogs. Today these rodents are much rarer. They have been exterminated in many places, partly because the holes they make can injure the domesticated livestock that graze throughout their grassland habitat. They can also be pests of cereal crops.

Eastern American chipmunk (*Tamias striatus*): 13–19cm (5–7.5in); 70–140g (0.15–0.2lb)
These rodents live in the eastern part of North America. They live in woodland and bushy habitat, feeding on nuts, seeds, mushrooms and fruit during the daytime. Eastern American chipmunks retire to burrows at night. They are solitary animals, sleeping through the winter in their burrows, waking frequently to feed on caches of food made during the autumn. Chipmunks carry spare food in cheek pouches.

American beaver

Castor canadensis

Beavers are among the largest of all rodents. Family groups of beavers live in large nests, called lodges, in or near forest streams or small lakes. Beavers eat wood and other tough plant foods, which have to be soaked in water before being eaten.

They use their large front teeth to gnaw through the bases of small trees. Sections of these logs are transported back to the lodge via a system of canals dug into the forest. If necessary, beavers will also dam a stream with debris to make a pool deep enough to store their food. A beaver colony may maintain a dam for several generations. The lodge has underwater entrances so beavers can swim out to their food supply even when the pool is frozen.

A beaver has webbed hind feet, a flattened tail for swimming and large front teeth for gnawing through wood. Its fur is coated with oil to keep it waterproof.

Distribution: North America.
Habitat: Streams and small lakes.
Food: Wood, leaves, roots and bark.
Size: 60–80cm (24–32in); 12–25kg (26–55lb).
Maturity: 1.5–2 years.
Breeding: 2–4 young born each spring.
Life span: 24 years.
Status: Locally common.

OTHER RODENTS

As well as squirrels, the Rodentia *order includes rats, mice and other small mammals, such as guinea pigs. Rodents are such a successful group partly because of their dentistry. They have large spaces between their incisor teeth at the front and the molars farther back, so that they can suck their cheeks in and close off their throats. This allows them to gnaw through wood or soil without swallowing it.*

North American porcupine

Erethizon dorsatum

North American porcupines are nocturnal animals that spend most of the night looking for food on the ground. However, they occasionally climb slowly into trees to find food. They cannot see very well, but have sensitive noses for detecting danger.

During the daytime, porcupines rest in hollow trees, caves or disused burrows. They regularly move from den to den throughout the year. They do not hibernate and keep feeding throughout the winter, but they will stay in their den during periods of harsh weather.

Porcupines live solitary lives, but do not defend territories, although they may drive away other porcupines from trees laden with food. If cornered by predators, porcupines turn their backs on their attackers and thrash around with their spiky tails. The barbed quills penetrate the attacker's skin and work their way into the body.

In early winter, males seek out females and shower them with urine before mating. The males are chased away by the females after mating. They give birth to their litters in summer.

Porcupines have sharp, barbed quills (thickened hairs) on their rumps and short tails for use in defence.

Distribution: North America, including Alaska and Canada, south to northern Mexico.
Habitat: Forest and brush.
Food: Wood, bark and needles in winter; buds, roots, seeds and leaves in summer.
Size: 64–80cm (25–32in); 3.5–7kg (7.5–15lb).
Maturity: 2.5 years.
Breeding: Single young born in summer.
Life span: 18 years.
Status: Common.

Capybara

Hydrochaerus hydrochaeris

Capybaras are the largest rodents in the world. They live in herds of about 20 individuals, feeding by day on the banks of rivers and in swampy areas. Although they are well suited to being in water, with eyes and nostrils high on the head and webbed hind feet, capybaras do not feed for long periods in water. They tend to use water as a refuge from predators and as a means of keeping cool on hot days. If startled, capybaras gallop into water and may swim to the safety of floating plants. When they surface, only their eyes and nostrils are visible.

Capybaras do not have permanent dens, but sleep in waterside thickets. Each herd contains several adults of both sexes as well as their offspring, all conforming to a hierarchy. A single male leads the herd. Only he can mate with the females in the herd. Fights often break out between the other males as they attempt to improve their rank.

Capybaras have bodies similar to guinea pigs, except that they are much bigger and more heavy-set. The males possess large sebaceous (oil) glands on their short rounded snouts.

Distribution: Central America to Uruguay.
Habitat: Thickly vegetated areas around fresh water.
Food: Grass, grains, melons and squashes.
Size: 1–1.3m (3.25–4.25ft); 27–79kg (59–174lb).
Maturity: 15 months.
Breeding: 5 offspring born throughout the year.
Life span: 10 years.
Status: Common.

Pygmy mouse

Baiomys taylori

Pygmy mice are the smallest rodents in the Americas, little more than the size of a person's thumb. They live in areas where plants, logs and rocks provide them with plenty of cover. The mice create networks of runs among undergrowth and under rocks, leaving piles of droppings at junctions. These may act as signposts or be signals to other mice in their network. Pygmy mice are most active at dawn and dusk, but will also feed throughout the day.

At night they sleep in nests made from plants and twigs. They do not live in groups as such, but will tolerate the presence of other mice close by. Pygmy mice can breed at a young age. Females can become pregnant after just a month of life. They breed throughout the year, often producing several litters per year. Both parents care for the young, which are born in nests inside shallow dips dug into the ground or in secluded cavities under logs or rocks.

The pygmy mouse's ears are smaller and rounder than most mice. It has black and brown hairs on its back with lighter red and brown fur underneath.

Distribution: South-western United States to central Mexico.
Habitat: Dry scrub.
Food: Stems, leaves, insects and seeds.
Size: 5–8cm (2–3in); 7–8g (0.015–0.017lb).
Maturity: Females 28 days; males 80 days.
Breeding: Several litters of 1–5 young each year.
Life span: 2 years.
Status: Common.

Mara (*Dolichotis patagonum*): 73–80cm (29–32in); 8–16kg (18–36lb)
The mara is a large rodent with long, thin legs, which make it look rather like a small deer. This species lives in the pampas of central and southern Argentina. Maras prefer to live on open grassland or in dry scrub. They dig their own burrows, using sharp claws on their forefeet. They stay in their burrows throughout the night and browse on plants throughout the day. Maras may travel in small groups, although they separate at night.

Chinchilla (*Chinchilla brevicaudata*): 22–38cm (9–15in); 500–800g (1.1–1.8lb)
Chinchillas are rare rodents that live in the Andes Mountains, ranging between Peru, Bolivia and Chile. Chinchillas are hunted illegally for their thick and soft blue-grey fur. They live in barren uplands, sheltering in crevices among rocks. They are active at dawn and dusk, eating any alpine plants that grow around them. Where they exist in large enough numbers, they live in colonies of up to 100 individuals. They produce two litters of two or three young each year.

Muskrat (*Ondatra zibethicus*): 22–33cm (9–13in); 0.7–1.8kg (1.5–4lb)
Muskrats live over most of North America, excluding the far north, and they are also found across Europe and warmer parts of Siberia. They are semi-aquatic animals, with oily, waterproof fur, webbed hind feet and scaly tails, used for steering while swimming. They live in small groups along riverbanks and in marshes. Muskrats are nocturnal, eating water plants and small aquatic animals. They dig burrows into riverbanks, which they access through underwater entrances, or they make domed nests from grass when living in more open wetlands.

Guinea pig

Cavia aperea

Guinea pigs, or cavies as they are also known, are most active in the twilight hours – around dawn and dusk. At these times, most predators are less active because their eyes cannot cope with the rapidly changing light levels. Guinea pigs are found in a wide range of habitats and altitudes, even living high up in the Andes Mountains.

Guinea pigs generally rest underground or in thickets. They may dig their own burrows, but are more likely to take over holes made by other animals. When on the move, these rodents follow well-trodden paths to areas where food is available.

Guinea pigs live in small groups of fewer than ten individuals. Many groups may crowd around a large supply of food, forming a temporary mass of rodents. The groups have hierarchies, with single males and females ruling over the others. Contenders for the top positions in the group may fight each other to the death. Breeding takes place throughout the year.

Although domestic guinea pigs often have long, soft coats of many colors and patterns, wild specimens have shorter and coarser fur, generally made up of grey, brown and black hairs.

Distribution: Colombia to Argentina, excluding the Amazon Basin.
Habitat: Grassland, swamp and rocky areas.
Food: Plants.
Size: 20–40cm (8–16in); 0.5–1.5kg (1–3.5lb).
Maturity: 3 months.
Breeding: Up to 5 litters per year.
Life span: 8 years.
Status: Common.

BATS

Bats are grouped together in the Chiroptera *order of mammals. They are the only mammals that can truly fly. Their wings are made from thin membranes of skin stretched between elongated arms and legs. Most bats are active at night and "see" the world through sound. They emit high-pitched calls and interpret the echoes that bounce back to build up pictures of their surroundings.*

Common vampire bat

Desmodus rotundus

Within their range, vampire bats are found in most types of habitat where there are large animals to feed upon, and they have become common in areas where livestock is being raised. They feed on the blood of animals such as cattle and donkeys and sometimes domestic poultry.

They begin to feed soon after nightfall, flying silently from their roosts in caves and hollow trees. Vampire bats will travel several kilometres to find blood. Once they locate suitable host animals, they lick the target area – usually on the neck or leg – and bite off hairs or feathers to clear a patch of skin. Then the bat cuts away a circle of skin with its long teeth and laps up the blood flowing from the wound.

Vampire bats swallow about 20ml (7fl oz) of blood each day. They return to their roosts to digest their food during the day. Roosts may contain as many as 2,000 bats. Single males mate with small groups of females. They need to guard them, however, because they are often usurped by other males. Births of single young take place in spring or autumn.

Vampire bats have dark upper bodies with grey undersides. Their upper front teeth are very long and pointed, and their limbs are adapted for walking along the ground.

Distribution: Mexico to Uruguay.
Habitat: Caves, hollow trees and disused buildings.
Food: Blood.
Size: 7–9cm (3–3.5in); 15–50g (0.03–0.1lb).
Maturity: 10 months.
Breeding: Single birth in spring or autumn.
Life span: 10 years.
Status: Common.

Tent-building bat

Uroderma bilobatum

Tent-building bats have four white stripes on their faces, with pointed "nose leaves". Up to 20 females may share a tent.

Tent-building bats live in areas with enough palm or banana trees for them to roost in. They make tents from the broad fronds by nibbling through the central, supportive ribs so that the fronds flop down over them. The tents shelter the bats from the sun and wind while they sleep during the day. The fronds eventually die and fall off the trees because their vascular systems have been damaged by the bats. Consequently, the bats build themselves new shelters every two or three months.

Tent-building bats eat mainly fruit, which they chew up, drinking the juice. They also alight on flowers to grab insects, and will eat any nectar and pollen available. Males roost alone or in small groups, while females rest in groups of 20 or more. Breeding takes place at all times of the year. Nursing mothers leave their single young in their tents while they go on their nightly foraging trips.

Distribution: Southern Mexico and northern South America to Brazil.
Habitat: Palm or banana forests.
Food: Fruit, pollen, nectar and insects.
Size: 5.5–7.5cm (2–3in); 13–21g (0.03–0.04lb).
Maturity: Not known.
Breeding: Single young born throughout the year.
Life span: Not known.
Status: Common.

Velvety free-tailed bat

Molossus ater

Velvety free-tailed bats are nocturnal insect-eaters, tracking their prey by echolocation. Echolocation is a system in which the bats bounce chirps of ultrasound off objects and listen to the echoes to build an image of their surroundings.

They live in damp forests, but will venture out into more open country to find food. They roost by day in tree hollows, in rock overhangs or under palm fronds. At dusk, the bats set off in search of food, which they store in pouches inside their cheeks. When the pouches are full, the free-tailed bats return to their roosts to digest their food.

These bats sometimes use their mobile tails as feelers by crawling backwards along the ground, waggling their tails from side to side. With wings adapted for twisting and turning in pursuit of prey, free-tailed bats are not very good at taking off from the ground. Instead, they take to the wing by climbing up trees and dropping into the air.

Velvety free-tailed bats are so named because of their soft fur and because, unlike most bats, they do not have membranes of skin joined to the sides of their tails.

Distribution: Northern Mexico to Argentina in South America.
Habitat: Forests and open woodland.
Food: Insects.
Size: 5–9.5cm (2–4in); 10–30g (0.02–0.04lb).
Maturity: 1 year.
Breeding: Single offspring produced up to 2 times per year.
Life span: Not known.
Status: Common.

Ghost bat (*Diclidurus albus*): 5–8cm (2–3in); 20–35g (0.04–0.07lb)
This species of bat ranges from southern Mexico to Peru and northern Brazil. Ghost bats have white or grey fur and are found in tropical forests, seldom far from running water. They roost by day under large palm leaves, and pursue moths and other insects by night. Ghost bats live largely solitary lives, but do congregate at roosts, especially during the breeding season in late summer.

Fishing bulldog bat (*Noctilio leporinus*): 10–13cm (4–5in); 60–80g (0.13–0.17lb)
The males of this species of bat, which lives in Central America, have bright orange fur, while the females are dull grey or brown. They have pointed muzzles with heavily folded lips and long hind legs with well-developed claws. These claws are used for catching fish. Fishing bulldog bats hunt over ocean surf as well as lakes and rivers. They even follow flocks of pelicans and snatch small fish disturbed by the birds.

New World sucker-footed bat (*Thyroptera discifera*): 3.5–5cm (1.5–2in); 40–60g (0.08–0.13lb)
This species ranges from Nicaragua to Peru and northern Brazil. Its name refers to the suction cups located on short stalks on the soles of its forefeet. The bats use these suckers to hang from smooth leaves in their rainforest habitat. Unusually for bats, which generally hang upside down, sucker-footed bats roost upright.

Pallid bat

Antrozous pallidus

Pallid bats prefer to live in areas with plenty of rocky outcrops, in dry scrubland or forest terrain in western North America. They roost in caves and hollow trees during the day and do not emerge until well after dark. They go on two foraging trips each night, returning to their roosts in between to digest their food. They hunt for food on the wing, frequently descending to about 2m (6.5ft) above the ground before taking a long glide over the terrain. This behaviour is suited to locating slow-moving and ground-based prey, such as beetles and crickets.

Some pallid bats may migrate from cooler parts of their range to warmer areas in winter. Others hibernate during the coldest months. Pallid bats live in large social groups. They call to one another as they return to the roosts after feeding, and communicate as they jostle for position inside their roosting sites. During the summer, males live in male-only roosts. Mating takes place in autumn, soon after that year's young have dispersed from their mothers' roosts. Births, usually of twins, take place in summer.

Pallid bats have creamy to yellow fur, with whitish patches on their undersides. Their ears are very large in proportion to their heads.

Distribution: Western North America from British Colombia to Mexico.
Habitat: Forests and arid scrubland.
Food: Insects, spiders and lizards.
Size: 6–8.5cm (2.5–3.5in); 17–28g (0.04–0.06lb).
Maturity: 1 year.
Breeding: Twins born in summer.
Life span: 9 years.
Status: Vulnerable.

HOOFED ANIMALS

Hoofed animals walk on the tips of their toes. Their hooves are made from the same material as fingernails and claws – keratin. Walking on tiptoes makes their legs very long, and most hoofed animals are fast runners because of this. Hoofed mammals belong to two groups: Perissodactyla *includes horses, zebras, rhinoceroses and tapirs, while* Artiodactyla *includes pigs, sheep, antelope, deer and cattle.*

Brazilian tapir

Tapirus terrestris

Brazilian tapirs, also known as South American tapirs, spend the day in forests of dense vegetation. At night they emerge into more open country, where they browse on vegetation. They prefer to spend part of the night in water or mud, and are surprisingly agile swimmers given their size. When on land, they walk with their snouts close to the ground. Each night, a tapir will follow one of several well-trodden trails to a favored watering hole.

Tapirs spend most of their lives alone. They are fairly aggressive towards one another at chance meetings. They alert each other to their presence by giving shrill whistling sounds and marking the ground with their urine. Brazilian tapirs breed all year round, but most mate during the rainy season, which means that their young are born just before the rains begin in the following year.

Tapirs have rounded bodies that are wider at the back than at the front. This helps them charge through thick vegetation when in danger. They have short hairs on their bodies and narrow manes on their necks. Their noses are long and flexible. Young tapirs have red fur patterned with yellow and white stripes and spots.

Distribution: From Colombia and Venezuela southward to northern Argentina.
Habitat: Woodland or dense grassy habitats with a source of water.
Food: Water plants, fruit and buds.
Size: 1.8–2.5m (6–8.25ft); 180–320kg (396–704lb).
Maturity: 3–4 years.
Breeding: 1 or 2 young born in rainy season.
Life span: 35 years.
Status: Lower risk.

Guanaco

Lama guanicoe

Guanacos have long limbs and necks for reaching food in trees and shrubs. They have brown, woolly fur on their upper bodies and necks, while their undersides have white hair.

Guanacos are considered the wild relatives of domestic llamas and alpacas. They are distant cousins of the camels of Africa and Asia. Like camels, guanacos have adapted to living in dry areas, although their preferred habitat – alpine grassland – is not as hot as the habitats of most camels.

Like their domestic relatives, guanacos are fast runners. They have more haemoglobin (oxygen-carrying pigment) in their red blood cells than any other mammal. This allows them to survive at altitude.

Guanacos mainly graze on grass, but they also pluck leaves from shrubs. They live in herds of about 15 individuals. Each herd is controlled by one adult male. Once a young guanaco reaches adulthood, it is chased away by the dominant male.

Distribution: Southern Peru to eastern Argentina and southern Chile.
Habitat: Dry, open areas.
Food: Grass.
Size: 1.2–2.5m (4–8.25ft); 100–120kg (220–264lb).
Maturity: Females 2 years; males 4 years.
Breeding: Single young born in spring.
Life span: 28 years.
Status: Vulnerable.

Bighorn sheep

Ovis canadensis

Bighorn sheep are named after the massive spiral horns of the males, which may be up to 1.1m (3.5ft) long. The females have smaller, less curved horns.

Bighorn sheep are the most common type of wild sheep in North America. They are excellent climbers and are often found on steep rocky outcrops or high cliffs. They seek refuge in steep areas from predators, such as cougars, which are not agile enough to keep up with their sure-footed prey.

Flocks of bighorn sheep can contain up to 100 individuals. They head up to high meadows in summer before retreating to the valleys when the winter snows come. Male bighorn sheep generally stay together in separate groups from the ewes and lambs. The rams have strict hierarchies based on the size of their horns. Fights are highly ritualized, with the adversaries butting their horns together. Ewes prefer to mate with rams with large horns and will refuse the courtship of others.

Distribution: South-western Canada to northern Mexico.
Habitat: Alpine meadows and rocky cliffs.
Food: Grass and sedge.
Size: 1.2–1.8m (4–6ft); 50–125kg (110–275lb).
Maturity: 3 years.
Breeding: 1–3 young born in spring.
Life span: 20 years.
Status: Lower risk.

Collared peccary (*Tayassu tajacu*): 0.75–1m (2.5–3.25ft); 14–30kg (31–66lb)
Collared peccaries, or javelines as they are also known, are relatives of wild boars and pigs. They have white collars around their necks and the rest of their bodies are covered in grey fur. They range from the south-western United States to northern Argentina. Peccaries root out bulbs and tubers with their snouts and also eat snakes and small invertebrates. Like many pigs, they appear to be immune to rattlesnake bites.

Muskox (*Ovibos moschatus*): 1.9–2.3m (6.25–7.5ft); 200–410kg (440–900lb)
Muskoxen are large cattle-like animals that live on the barren tundra of the Canadian Arctic. Males are larger than females, and both sexes have curved horns and coats of thick, dark brown hair that hang almost to the ground. Their name is derived from the strong smell of the bulls during the mating season in summer.

Vicuña (*Vicugna vicugna*): 1.25–1.9m (4–6.25ft); 36–65kg (79–143lb)
Vicuñas are related to guanacos and camels. They live in the high Andes Mountains of Peru, Bolivia and Chile. They are smaller than guanacos and have tawny colored coats. Some biologists have suggested that alpacas are a cross between domestic llamas and vicuñas, although this theory is contestable. Vicuñas have teeth more like those of rodents than other hoofed animals. Their lower incisors grow throughout the animal's lifetime. The teeth are constantly being worn away by the tough alpine grasses that make up the vicuña's diet.

American bison

Bison bison

Although rare, the American bison has been saved from extinction. Once, herds of over a million individuals grazed the vast prairies of western North America. They were almost wiped out by hunters in the 19th century where the grasslands were cleared to make way for agriculture.

Bison often rub their shoulders and rumps against boulders and tree trunks, and they enjoy taking mud and dust baths. This helps them scratch off fly larvae and the other parasites that live on their hides.

Mature bulls move in separate groups from the cows. In the mating season, in late summer, the males join the females. They fight for the females by ramming each other head-on. After mating, a bull will guard his mate for several days to prevent other rival males from mating with her.

Male bison are larger than females. Both sexes have sharp, curved horns, which stick out from the shaggy, brown hair on their heads.

Distribution: Patches of western Canada and central United States.
Habitat: Prairie and woodland.
Food: Grass.
Size: 2.1–3.5m (7–11.5ft); 350–1000kg (770–2200lb).
Maturity: 1–2 years.
Breeding: Single young born in spring every 1 or 2 years.
Life span: 40 years.
Status: Lower risk.

DEER

Deer are a group of hoofed mammals that are found across the northern hemisphere. They belong to the
Cervidae family of mammals. In form and habit, deer resemble the horned antelopes of Africa, which are
actually more closely related to sheep and cattle. However, instead of horns, deer grow antlers. In most
species, only males have them, and unlike horns, which remain for life, antlers are shed annually.

Moose

Alces alces

Male moose are almost twice the size of females. The males sport huge antlers – nearly 2m (6.5ft) across – and have flaps of skin hanging below their chins, called dewlaps.

Moose are the largest deer in the world.
They live in the cold conifer forests
that cover northern mountains and
lowlands. As well as in North
America, moose live across northern
Europe and Siberia, but are known
as elk there.

Moose plod through the forests and
marshes, browsing on a wide range of
leaves, mosses and lichens. They often
feed in the shallows of streams and rivers,
nibbling on aquatic vegetation. These large deer
have even been seen diving underwater to uproot
water plants. In summer, they are most active at dawn
and dusk. In winter, they are active throughout the day.
They paw the snow to reveal buried plants and twigs.
Although moose may gather together
to feed, they spend most of the year
alone. In the autumn mating season,
males fight each other for females.

Distribution: Alaska, Canada,
northern United States, Siberia
and northern Europe.
Introduced to New Zealand.
Habitat: Marsh and
Woodland.
Food: Leaves, twigs, moss
and water plants.
Size: 2.4–3.1m (8–10.5ft);
200–825kg (440–1815lb).
Maturity: 1 year.
Breeding: 1–3 young born
in spring.
Life span: 27 years.
Status: Common.

White-tailed deer

Odocoileus virginianus

White-tailed deer, or Virginia deer as they are called in the
United States, prefer areas with tall grasses or shrubs to hide
in during the day. When the deer spot predators, they raise
their white tails to expose the white patches on their rumps.
This serves as a visual warning to other deer that danger is
near. If pursued, the deer bound away,
reaching 60kph (40mph).

White-tailed deer live in
matriarchies, with each small group
being controlled by a single adult
female, which is the mother of the rest
of the group. The adult males live alone
or in small bachelor herds. In the autumn
mating season, males mark plants with
scent produced by glands on their faces,
and urinate in depressions scraped into
the ground. The males fight with
their antlers – rut – for the
right to court females.

*White-tailed
deer have
brown fur
on their upper
parts and white
undersides. The
white fur extends under
the tail, which gives the
species its name. The
males shed their antlers
in midwinter and grow
new ones in spring.*

Distribution: North, Central
and South America from
southern Canada to Brazil.
Habitat: Shrublands and
open woodland.
Food: Grass, shrubs,
mushrooms, twigs, lichens
and nuts.
Size: 0.8–2.1m (2.5–7ft);
50–200kg (110–484lb).
Maturity: 1 year.
Breeding: 1–4 young born
in summer.
Life span: 10 years.
Status: Common.

Pronghorn

Antilocapra americana

Despite appearances, pronghorns are not true deer. They are the sole members of a separate group of hoofed animal called the *Antilocapridae*. Unlike true deer, pronghorns do not have antlers, but have horns like antelope, although they are forked like those of a deer.

Pronghorns are the fastest land mammals in the Americas. They have been recorded racing along at 72kph (45mph).

In late autumn, pronghorns gather into large herds of 1,000 or more. They spend the winter in these herds and split into smaller single-sex groups when spring arrives. In October, older males compete for small territories, which they use to attract groups of females. Once females have entered the territories, the males will not allow other males near them.

Pronghorns get their name from the prongs sticking out halfway up their backward-curving horns. Male pronghorns are slightly larger than the females. The males also have black masks on their faces.

Distribution: Southern Canada to northern Mexico.
Habitat: Grassland and desert.
Food: Grass, leaves and cacti.
Size: 1–1.5m (3.25–5ft); 36–70kg (79–154lb).
Maturity: Females 18 months; males 3 years.
Breeding: 1–3 young born in spring.
Life span: 10 years.
Status: Common.

Red brocket deer (*Mazama americana*): 72–140cm (27–54in); 8–25kg (17–55lb)
Red brocket deer range from eastern Mexico to northern Argentina. They have whorls of hair on their faces and stout bodies covered in reddish-brown hair. The males have simple spike-like antlers. This species lives in woodland and dense forests. They may be active at all times of the day or night, feeding on grasses, vines and the new shoots of plants. These shy deer tend to freeze when they spot danger, blending into the thick vegetation.

Pampas deer (*Ozotoceros bezoarticus*): 1.1–1.4m (3.5–4.75ft); 25–40kg (55–88lb)
As their name suggests, pampas deer live on the open grasslands – or pampas plains – in the south-eastern part of South America. They have dark red or brown coats. Males have forked antlers and glands on their hooves that produce a strong scent. Pampas deer graze on young grass shoots throughout the day. In late summer and autumn the males fight for access to females, and most births occur in the spring.

Huemul (*Hippocamelus antisensis*): 1.4–1.65m (4.75–5.5ft); 45–65kg (99–143lb)
Huemuls live in the rugged hill country high in the Andes of Peru, Chile, Bolivia and Argentina. They have coarse coats and black Y-shaped markings on their faces. Huemuls spend the summer high on the mountains, grazing on the grasses and sedges growing in alpine meadows. In winter, they climb down to lower altitudes. Mating takes place in the drier winter season and fawns are born at the end of the rains.

Southern pudu

Pudu pudu

Southern pudus live in the wet forests on the slopes of southern Andes Mountains. They are the smallest of all deer, being only a little bigger than maras – the long-legged rodents that live in the same region.

Pudus are mainly nocturnal in the way they behave, although they are sometimes spotted feeding during the day. They move through the dense forest slowly, picking off the ripest fruits and most succulent leaves and buds. They try not to draw attention to themselves and stay well hidden as much as possible. If attacked, they run away in zigzag paths and often seek refuge in the branches of trees.

Pudus live alone, patrolling small territories and only occasionally encountering other members of their species. Being so small, pudus can reach maturity much more quickly than other deer. They are ready to breed after just six months of life. Births take place in spring, and by the following autumn that year's fawns are ready to take part in the breeding activity, which occurs at that time.

Southern pudus have grey and brown fur. They have short, thick legs and the males have small spikes for antlers.

Distribution: Southern Chile to south-western Argentina.
Habitat: Humid forest.
Food: Tree and shrub leaves, vines, bark, fruit and flowers.
Size: 60–85cm (23–33in); 5–14kg (11–30lb).
Maturity: 6 months.
Breeding: Single young in spring.
Life span: 10 years.
Status: Vulnerable.

NEW WORLD MONKEYS

These monkeys are so named because the Americas comprise the New World. They are the only primates that live in the Americas, and most live in South America, with only a few species ranging as far north as Central or Meso America and Mexico. New World monkeys differ from Old World monkeys in two ways: they have flattened noses with more broadly spaced nostrils, and most have prehensile tails.

Pygmy marmoset

Cebuella pygmaea

Distribution: Upper Amazon Basin.
Habitat: Rainforest.
Food: Fruit, buds, insects and sap.
Size: 11–15cm (4.5–6in); 100–140g (0.22–0.3lb).
Maturity: 1.5–2 years.
Breeding: 1–3 young born throughout the year.
Life span: 10 years.
Status: Common.

Pygmy marmosets are the smallest monkeys in the world. They live in the low plants that grow beneath tall trees in tropical forests. They clamber among the thick vegetation in search of food throughout the day, being most active in the cooler hours at the beginning and end of each day.

Pygmy marmosets eat fruit, flower buds and insects, but their preferred food is the sweet, sticky sap from certain trees. Their lower canine teeth are specially shaped for gouging holes in tree bark, causing the sap to leak out from the wood beneath. A tree used by a group of pygmy marmosets will be covered in wounds where the animals repeatedly bite through the bark.

Like all monkeys, pygmy marmosets live in complex societies. They live in family groups, with two parents and eight or nine offspring. Families sleep together, huddled on branches. Breeding pairs may mate at any time, but most produce small litters – usually twins – in December or June.

Pygmy marmosets have grey-brown fur and tails ringed with red-brown stripes. Their tails are prehensile and can be wrapped around branches. They move cautiously through the treetops to avoid attack by large birds of prey.

Brown capuchin

Cebus apella

Capuchins are sometimes called ring-tails. This is because they often curl the tip of their semi-prehensile tail into a ring. Capuchins are among the most intelligent and adaptable of all monkeys. They are found in a wide range of habitats, from dense jungles to towns and cities. Some even live on the seashore, where they collect crabs. Some capuchins break open hard nuts by pounding them with stones – an example of animals using tools.

This species lives in troops of about 12 monkeys. Most troops have a single adult male, who fathers all the children. The monkeys chatter and squeak a great deal, telling each other of their location and warning of danger. Without a set breeding season, capuchin mothers give birth to their single babies at any time of the year. Each baby initially clings to its mother's chest, then rides on her back until it becomes more independent.

Like some other capuchin species, brown capuchins have a cap of dark hair on the top of their heads, with thick tufts or "horns" above the ears.

Distribution: From Colombia to Paraguay.
Habitat: Rainforest.
Food: Fruit, nuts, flowers, bark, gums, insects and eggs.
Size: 30–56cm (12–22in); 1.1–3.3kg (2.5–7.25lb).
Maturity: Females 4 years; males 8 years.
Breeding: Single young born throughout the year.
Life span: 30 years.
Status: Lower risk.

Humboldt's woolly monkey (*Lagothrix lagotricha*): 50–69cm (20–27in); 5.5–11kg (12–24lb)
Woolly monkeys live in the high forests on the slopes of Andes Mountains and Mato Grosso in Brazil. They have very heavyset bodies compared to other tree-living monkeys, with thick black fur. They spend much time at the tops of the tallest trees that protrude from forest canopies, although they do climb down to the ground more often than most monkeys. On the ground, they may walk on their hind legs using their heavy tails to keep them upright. They eat mainly fruit and insects, but can survive on leaves and seeds when necessary.

Golden lion tamarin (*Leontopithecus rosalia*): 20–33cm (8–13in); 600–800g (1.3–1.7lb)
This extremely rare monkey lives in just one small area of coastal forest near Rio de Janeiro, Brazil. The long, silky, golden mane around the head and shoulders has earned the golden lion tamarin its English name. These monkeys feed on fruit, insects and small vertebrates during the day, before retiring to holes in trees to sleep. They live in family groups of four or five, the male helping to raise his offspring. Golden lion tamarins breed during the wettest part of the year, which is during the warm southern summer. Most offspring are born as twins.

Squirrel monkey

Saimiri boliviensis

Squirrel monkeys live in many types of forest. They spend most of their time in trees, rarely coming down to the ground. However, some populations of squirrel monkeys have made their homes in areas cleared of trees for agriculture. These monkeys tend to live close to streams for reasons of safety. Squirrel monkeys form complex social groups, or troops, which are larger than those of any other monkey species in the Americas. In pristine rainforests, the troops can number up to 300 individuals.

Males do not help in raising the young and during the mating season – the dry part of the year – they establish hierarchies by fighting each other. Only the dominant males get to mate with the females, which give birth to single young. Soon after giving birth, the new mothers chase away the breeding males, which reform their bachelor subgroups. Adolescent males, too old to stay with their mothers, eventually join these subgroups, having fought their way in.

Distribution: Central America to Upper Amazon Basin.
Habitat: Tropical forest, close to streams.
Food: Fruit, nuts, flowers, leaves, gums, insects and small vertebrates.
Size: 26–36cm (10–14in); 0.75–1.1kg (1.65–2.4lb).
Maturity: Females 3 years; males 5 years.
Breeding: Single young born from June–August.
Life span: 20 years.
Status: Lower risk.

Squirrel monkeys have black, hairless snouts and helmets of dark fur around their pale faces. Their ears are covered in pale fur. The rest of the body is more brightly colored, in hues of pale yellow and red, and the mobile tail has a black tip. The body is slender in shape.

Black-handed spider monkey

Ateles geoffroyi

Spider monkeys are the most agile of the American primates, not least because their long prehensile tails function as a fifth limb. The animals can pick up food or hold on to branches with their tails. It is not unusual to see one of these monkeys hanging from its tail alone.

Spider monkey troops live high up in forest canopies, and almost never visit the ground. They are most active early in the morning, taking the rest of the day to relax and digest tough plant food.

Troops generally contain about 30 individuals, with both sexes in equal numbers. However, larger groups of more than 100 have been reported. The males in the troops defend large territories by regularly patrolling the perimeters, while females and young tend to stay close to the centre. Males tend to stay in the troops they were born into, while females move to other troops in the area. Breeding occurs all year round. Spider monkeys reportedly have a unique defensive strategy: when potential predators approach – including humans – the monkeys drop heavy branches on top of them.

Spider monkeys have very long prehensile tails and similarly long legs, hence their name. This allows them to be extremely agile in the treetops.

Distribution: Mexico to Colombia.
Habitat: Tropical forest.
Food: Fruit, seeds, buds, leaves, insects and eggs.
Size: 38–63cm (15–25in); 6–8kg (13.2–17.6lb).
Maturity: 4–5 years.
Breeding: Single young born throughout the year.
Life span: 30 years.
Status: Vulnerable.

Red howler monkey

Alouatta seniculus

Howlers are large monkeys that live in the trees of tropical forests. They are known for the roaring howls that fill South American forests. The monkeys have very wide jaws, which allow them to open their mouths wide and make such loud calls.

Howler monkeys roar first thing in the morning before setting off to look for food. Although they eat fruits, such as figs, when they are available, howler monkeys rely for long periods on just leaves. Few other monkeys have such an unvaried, indigestible diet. After a rest in the middle of the day, the monkeys feed some more before travelling back to their sleeping trees while howling to each other again.

Howler monkey troops contain about eight or nine individuals. Larger troops form when there are more fruits available. Males compete with each other to join troops, and the victors may kill the young of the males they depose. The howling call is thought to be a mechanism for locating nearby troops. These monkeys breed all year round. The young ride on their mothers' backs for up to a year. Both males and females leave their mothers' troops when they are sexually mature and join others.

Distribution: Northern South America.
Habitat: Rainforest and mangroves.
Food: Leaves and figs.
Size: 55–92cm (22–36in); 4–10kg (8.75–22lb).
Maturity: 4–5 years.
Breeding: Single young born throughout the year.
Life span: 20 years.
Status: Endangered.

Howler monkeys typically have reddish-brown hair, although some have a more yellowish or dusky coloration. Their strong prehensile tails have naked patches on their undersides to help them grip branches. The males are larger than the females and generally have darker hair. The loud calls of these monkeys, especially by the males, are made possible by a specialized larynx in the throat, which amplifies the sound.

White-faced saki

Pithecia pithecia

Distribution: Northern South America from Venezuela to north-eastern Brazil.
Habitat: Tropical forest.
Food: Fruit, honey, leaves, mice, bats and birds.
Size: 30–70cm (12 28in); 0.7–1.7kg (1.5–3.75lb).
Maturity: 2–3 years.
Breeding: Single young born in the dry season remain with their parents until they are mature.
Life span: 14 years.
Status: Common.

White-faced sakis live high up in trees. They feed during the daytime and almost never come down to the ground. Although they do occasionally leap from tree to tree, sakis are not the most agile of monkeys. They climb down trunks backwards and generally run along thick branches on all fours. Sometimes, however, sakis have been seen walking on their hind legs with their arms held above their heads.

A lot of the saki's diet consists of vegetable matter and fruit. However, these monkeys do also catch small vertebrate animals, such as birds and bats. The sakis rip their victims apart with their hands before skinning and eating the pieces of flesh. The monkeys have sharp teeth that are useful for biting into forest fruits and slicing up meat.

A saki group is based around a pair of breeding adults. The rest of the group, which may contain up to five individuals, will generally be the chief pair's offspring of different ages. The breeding pair produce a single baby once a year. Most births occur in the dry season at the end of the year.

Only male white-faced sakis have the white faces after which they are named. The females have black or dark brown faces. Most saki monkeys have broad, round faces with hooded eyebrows.

Night monkey

Aotus lemurinus

Night monkeys, which are also known as douroucoulis, are the only nocturnal monkeys in the world. These rare monkeys live in most types of forest except those close to water. Biologists used to think there was a single species, but it is now known that there are several living across South America.

The large eyes of these monkeys collect enough light for them to see in the gloom. Night monkeys can only see in monochrome (black and white), but this still allows them to run and jump through the trees even on the darkest nights. By day, they rest in nests made from dry leaves and twigs.

Night monkeys live in family groups, with one adult pair and two or three of their young. Family members warn each other of approaching danger, such as tree snakes or birds of prey, with long "wook" alarm calls. The monkeys have loose sacs of skin under their chins, which they inflate to amplify these calls. At night these visual signals are ineffective, so night monkeys rely on scent as well as calls to communicate with other monkeys and nearby groups. The scent comes from their urine and glands on their chests, which the monkeys rub on branches.

Night monkeys have large eyes which give them good night vision. Their thick, woolly fur gives them a rounded appearance, and their tails, which are not prehensile, are thickened and furry at their tips.

Distribution: Central and South America from Panama southward to Brazil, but patchily distributed.
Habitat: Forests.
Food: Fruit, nuts, leaves, bark, flowers, gums, insects and small vertebrates.
Size: 24–37cm (9.5–14.5in); 0.6–1kg (1.25–2.25lb).
Maturity: 2 years.
Breeding: Single young born throughout the year.
Life span: 18 years.
Status: Vulnerable.

Titi monkey (*Callicebus molloch*): 24–61cm (9.5–24in); 0.5–0.75kg (1–1.75lb)
Titi monkeys live in most forested parts of northern South America. They exhibit a variety of fur colors, ranging from grey to red to gold. Their tails are long and bushy, but not prehensile. The monkeys tend to live in the lower trees near riverbanks. They often climb down to the shrub plants near to the ground to feed on fruit, leaves, birds' eggs, and invertebrates. They live in family groups dominated by single pairs of adults.

Black-bearded saki (*Chiropotes satanas*): 32–51cm (12.5–20in); 2–4kg (4.5–8.75lb)
As its name suggests, the black-bearded saki has a very long, thick beard on its elongated chin. Its head, beard and tail are black, and its shoulders, back, hands and feet are reddish-brown to black. This species lives in Guyana, Venezuela and Brazil, north of the Amazon. The black-bearded saki eats fruit, the seeds of unripe fruit, leaves, flowers and a few insects. It lives in troops of up to 30 individuals.

Muriqui (*Brachyteles arachnoides*): 46–63cm (18–25in); 12–15kg (44–33lb)
The muriqui, or woolly spider monkey, is the largest New World monkey. This extremely rare monkey lives in the coastal forests of south-eastern Brazil – an area which has been heavily deforested. Little is known about muriqui society. The animals are thought to live in promiscuous groups, where all adults freely mate with each other.

Bald uakari

Cacajao calvus

Uakaris only live in tropical rainforests that are flooded or filled with many slow-flowing streams, and consequently they are very rare. They are active during the day, running on all fours through the tops of large trees. They mainly feed on fruit, but will also eat leaves, insects and small vertebrates. Although they are quite agile, uakaris rarely jump from branch to branch. They almost never come down to the ground.

Uakaris live in large troops of 10–30 individuals. In areas where forests have not been damaged by human activity, groups of over 100 have been reported. Uakari troops often get mixed in with those of other monkeys, such as squirrel monkeys, during daytime feeding forays.

Each troop has a hierarchical structure, which is maintained by fighting among both sexes. The dominant males control access to females in a troop during the breeding season. Females give birth to single young every two years.

Distribution: Upper Amazon from Peru to Colombia.
Habitat: Beside rivers in flooded forests.
Food: Fruit, leaves and insects.
Size: 51–70cm (20–28in); 3.5–4kg (7.5–8.75lb).
Maturity: Females 3 years, males 6 years.
Breeding: Single young born in summer every 2 years.
Life span: 20 years.
Status: Endangered.

Bald uakaris have hairless, red faces fringed with shaggy fur, hence their name. The long fur on the body is pale but looks reddish-brown, and a few have white fur. Their clubbed tails are proportionally shorter than all other New World monkeys.

ARMADILLOS AND RELATIVES

Armadillos, anteaters and sloths belong to a group of mammals called the Xenarthra *(formerly named* Edentata, *meaning toothless). Most xenarths live in South and Central America. Only one, the long-nosed armadillo, lives as far north as Texas. These animals are taxonomically related to one another but do not share evident common physical characteristics, except for unique bones that strengthen their spines.*

Giant armadillo

Priodontes maximus

Like all armadillos, the giant armadillo has bands of bony plates running from side to side across its body to serve as armour. These plates are covered in leathery skin, and a few thick hairs stick out from between them.

The giant armadillo is the largest of all armadillos. It is nocturnal and shelters by day in burrows dug with the mighty claws on its forefeet. Most of the burrows are dug into the side of termite mounds. Giant armadillos also dig to get at their prey. They typically excavate termite mounds and ant nests, but they also dig out worms, subterranean spiders and occasionally snakes.

Unlike many other armadillos, giant armadillos cannot curl up completely to protect their soft undersides with their armoured upper bodies. Instead, these giants rely on their considerable size to deter predators. If they are attacked, giant armadillos try to dig themselves out of trouble. Armadillos live alone. They breed all year and mate when they chance upon one another on their travels. One or two young are born inside a large burrow after a four-month gestation.

Distribution: Venezuela to northern Argentina.
Habitat: Dense forest and grassland near water.
Food: Termites, ants, spiders and other insects, worms, snakes and carrion.
Size: 0.7–1m (2.25–3.25ft); 60kg (132lb).
Maturity: 1 year.
Breeding: 1–2 young born throughout the year.
Life span: 15 years.
Status: Vulnerable.

Long-nosed armadillo

Dasypus novemcinctus

Distribution: Southern United States to northern Argentina.
Habitat: Shaded areas.
Food: Arthropods, reptiles, amphibians, fruit and roots.
Size: 24–57cm (9.5–22.5in); 1–10kg (2.2–22lb).
Maturity: 1 year.
Breeding: 4 young born in spring.
Life span: 15 years.
Status: Common.

Long-nosed armadillos are found in a wide range of habitats, but always require plenty of cover. In the warmer parts of their range they feed at night. In colder areas they may be spotted during the day, especially in winter. These armadillos build large nests at the ends of their long burrows. The nests are filled with dried grasses. In areas with plenty of plant cover, long-nosed armadillos may also build their nests above ground.

Long-nosed armadillos search for their animal prey by poking their long noses into crevices and under logs. They also eat fallen fruit and roots. When threatened, the animals waddle to their burrows as fast as possible. If cornered, they will curl up into armoured balls.

Long-nosed armadillos forage alone, but they may share their burrows with several other individuals, all of the same sex. The breeding season is in late summer. Litters of identical same-sex quadruplets are born in the spring.

The long-nosed armadillo is also called the nine-banded armadillo because it typically has that number of plate bands along its back, although specimens can possess either eight or ten bands.

Three-toed sloth

Bradypus tridactylus

Distribution: Eastern Brazil.
Habitat: Coastal forest.
Food: Young leaves, twigs and buds.
Size: 41–70cm (16–29in); 2.25–5.5kg (5–12lb).
Maturity: 3.5 years.
Breeding: Single young born throughout the year.
Life span: 20 years.
Status: Endangered.

Three-toed sloths spend most of their lives hanging upside down from trees. They are very inactive creatures, but they do climb down to the ground once or twice a week to excrete or move to other trees.

Three-toed sloths feed by pulling on flimsy branches with their forelegs, to bring them close to their mouths. They spend long periods waiting for their tough food to digest. Because they are so inactive, sloths have a lower body temperature than other mammals – sometimes as low as 24°C (75°F). Their fur is sometimes tinged with green because algae are growing in it. The sloths may absorb some of the algal nutrients through their skin, and it also provides useful camouflage from predators.

Sloths have very simple societies. They live alone and females only produce offspring every two years. Mating can occur throughout the year though, with both partners still hanging upside down. Mothers give birth in this position too, and the young cling to the hair on their mothers' breasts.

Three-toed sloths have long grey hairs that project in the opposite direction to those of other mammals, so that they point downwards when the animals are upside down. This ensures that rainwater runs off. They climb using their strong, hook-like claws.

Pichi (*Zaedyus pichiy*): 26–34cm (10–13in); 1–2kg (2.2–4.4lb)
The pichi is a small armadillo that lives on the grasslands of southern Argentina and the alpine meadows of the Chilean Andes. Its head, body and tail are armoured and have long hairs growing out from behind each plate. When threatened, a pichi draws its legs under its body so that the serrated edges of its armour dig into the ground. It anchors itself in its burrow in this way. Pichis probably hibernate in colder parts of their range. Between one and three young may be born at any time of the year.

Northern tamandua (*Tamandua mexicana*): 47–77cm (18–30in); 2–7kg (4.4–15.5lb)
Tamanduas are sometimes called lesser anteaters. They range from southern Mexico to northern Peru. Their fur is short and thick, and varies in color. Tamanduas have long snouts with extremely small mouths, and long, curved claws. Lacking teeth – like all anteaters – their insect food is ground up using muscular gizzards inside their stomachs. They feed on ants, termites and bees, searching for them both on the ground and in trees.

Two-toed sloth (*Choloepus hoffmanni*): 54–74cm (21–29in); 4–8.5kg (8.8–19lb)
Two-toed sloths live in the rainforests of northern South America. Unlike their three-toed cousins they have only two digits on each forefoot, but their hind feet have three. Like other sloths, this species lives a slow, nocturnal life, hanging from high branches. It eats leaves, fruit and twigs. Algae grow on its long fur, camouflaging it from predators.

Giant anteater

Myrmecophaga tridactyla

Giant anteaters live wherever there are large ant nests or termite mounds in abundance. They use their powerful claws to rip the colonies apart, then they use their sticky tongues to lick up the insects and their eggs and larvae. A single giant anteater can eat over 30,000 ants or termites in one day.

Despite being powerful diggers, giant anteaters shelter in thickets, not burrows, because of their awkward shape. They spend most of their time alone searching for food, with their long noses close to the ground. While on the move, they curl their forelimbs under their bodies so that they are actually walking on the backs of their forefeet and their claws do not hinder them.

Females often come into contact with one another, but males keep their distance. Breeding can take place all year.

Distribution: Belize to northern Argentina.
Habitat: Grasslands, forests and swamps.
Food: Ants, termites and beetle larvae.
Size: 1–1.2m (3.25–4ft); 18–39kg (40–89lb).
Maturity: 2.5–4 years.
Breeding: Single young born throughout the year.
Life span: 25 years.
Status: Vulnerable.

Giant anteaters have powerful digging claws on their forelimbs and incredibly long tongues – often over 60cm (24in) – inside their snouts. They have white stripes along their flanks and a long, bushy tail.

RABBITS

Rabbits, hares and pikas belong to the mammal order Lagomorpha. *Most of the lagomorphs that live in the Americas are found north of Mexico. Unlike their cousins in Europe, most American rabbits do not dig burrows. Like hares, they generally shelter above ground. The only rabbit to dig its own burrow is the pygmy rabbit, which is also the smallest rabbit in the world.*

Pika

Ochotona princeps

The pikas of North America live in areas of scree – fragments of eroded rock found beneath cliffs or mountain slopes. They shelter under the rocks and feed on patches of vegetation that grow among the scree. Pikas may forage at all times of the day or night, but most activity takes place in the early mornings or evenings.

During the winter, pikas survive by eating "ladders" of grass and leaves that they have collected during the late summer. These ladders are piles made in sunny places, so that the plants desiccate into alpine hay. Like most of their rabbit relatives, pikas eat their primary droppings so that their tough food is digested twice, in order to extract all of the nutrients.

Adult pikas live alone and defend territories during the winter. In spring, males expand their territories to include those of neighbouring females. Most females produce two litters during each summer. When preparing for winter, the females chase their mates back to their own territories and expel their mature offspring.

Pikas are small relatives of rabbits. They do not have tails, and their rounded bodies are covered in soft red and grey fur. Unlike those of a rabbit, a pika's hind legs are about the same length as its forelegs.

Distribution: South-western Canada and western United States.
Habitat: Broken, rocky country and scree.
Food: Grass, sedge, weeds and leaves.
Size: 12–30cm (5–12in); 110–180g (0.25–0.4lb).
Maturity: 3 months.
Breeding: 2 litters born during summer.
Life span: 7 years.
Status: Common.

Snowshoe hare

Lepus americanus

Like most hares, snowshoe hares do not dig burrows. Instead they shelter in shallow depressions called forms, which they scrape in soil or snow. Snowshoe hares are generally nocturnal, and rest in secluded forms or under logs during the day. When dusk arrives, the hares follow systems of runways through the dense forest undergrowth to feeding sites. They maintain these runways by biting away branches that block the way and compacting the winter snow.

In summer, the hares nibble on grasses and other green plant material. They survive the long winter by supplementing their diet with buds, twigs and bark. Over several years, the overall population of snowshoe hares can rise and fall dramatically. At the low points there may be only two animals per square kilometre. At the peak there may be as many as 1,300 in the same area.

Snowshoe hares are more social than other hares. During the spring breeding season, the male hares compete with each other to establish hierarchies and gain access to mates. Conflicts often result in boxing fights – hence "mad March hares".

Distribution: Alaska, Canada and northern United States.
Habitat: Conifer forest.
Food: Grass, leaves, buds and bark.
Size: 40–70cm (16–27in); 1.35–7kg (3–15lb).
Maturity: 1 year.
Breeding: 4 litters per year.
Life span: 5 years.
Status: Common.

Female snowshoe hares are a little larger than the males. In summer their fur is grey, but in areas with heavy winter snow, they grow white fur as camouflage from predators.

Eastern cottontail

Sylvilagus floridanus

Cottontail rabbits do not dig burrows, although they may shelter in disused ones dug by other animals. Generally they shelter in thickets or forms – shallow depressions made in tall grass or scraped in the ground. Cottontails forage at night, grazing mainly on grasses, but also nibbling small shrubs. Unlike hares, which rely on their speed to outrun predators, cottontails freeze when under threat, blending into their surroundings. If they have to run, they follow zigzag paths, attempting to shake off their pursuers.

In warmer parts of their range cottontails breed all year round, but only in summer farther north. Males fight to establish hierarchies, with the top males getting their choice of mates. A pregnant female digs a shallow hole, which is deeper at one end than the other. She lines the nest with grass and fur from her belly. Once she has given birth, she crouches over the shallow end and her young crawl up from the warm deep end to suckle.

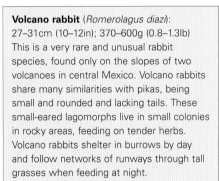

Females cottontails are larger than the males. The name "cottontail" is derived from their short rounded tails, which have white fur on their undersides. Their upper bodies are covered in grey, brown and reddish hairs.

Distribution: Eastern Canada and United States to Venezuela.
Habitat: Farmland, forest, desert, swamp and prairie.
Food: Grass, leaves, twigs and bark.
Size: 21–47cm (8–18in); 0.8–1.5kg (1.75–3.3lb).
Maturity: 80 days.
Breeding: 3–7 litters each year.
Life span: 5 years.
Status: Common.

Volcano rabbit (*Romerolagus diazi*): 27–31cm (10–12in); 370–600g (0.8–1.3lb) This is a very rare and unusual rabbit species, found only on the slopes of two volcanoes in central Mexico. Volcano rabbits share many similarities with pikas, being small and rounded and lacking tails. These small-eared lagomorphs live in small colonies in rocky areas, feeding on tender herbs. Volcano rabbits shelter in burrows by day and follow networks of runways through tall grasses when feeding at night.

Pygmy rabbit (*Brachylagus idahoensis*): 21–27cm (8–11in); 200–450g (0.4–0.9lb) These rabbits live on an arid plateau in the north-west of the United States. Pygmy rabbits are related to cottontails, but they are about half the size. They dig burrows under thickets of sagebrush – the only North American rabbit to do so – and move through a network of runways above ground. They eat the sagebrush and are most active at dawn and dusk.

Swamp rabbit (*Sylvilagus aquaticus*): 45–55cm (18–22in); 1.5–2.5kg (3.3–5.5lb) Swamp rabbits live in the wetlands around the Mississippi Delta and other rivers in the southern United States. Unlike most rabbits, the males and females of this species are about the same size. Swamp rabbits build nests of dead plants and fur at ground level. They maintain territories by calling to intruders and marking their areas with scent. They breed all year round. Female swamp rabbits may produce up to 40 young per year.

Jackrabbit

Lepus californicus

Jackrabbits are actually a type of hare and so share many of the hare's characteristics, from long ears to large, hairy hind feet. Jackrabbits live in dry areas with only sparse plant cover. This has benefited the species in the past. Overgrazing of the land by cattle in the arid south-west United States and Mexico has created an ideal habitat for jackrabbits.

Unlike other hares, jackrabbits make use of burrows. They do not dig their own, but they will modify underground shelters made by tortoises trying to escape the heat of the day.

Distribution: South-western United States to northern Mexico.
Habitat: Dry grasslands.
Food: Grass.
Size: 40–70cm (16–28in); 1.35–7kg (3–15lb).
Maturity: 1 year.
Breeding: 3–4 litters each year.
Life span: 5 years.
Status: Common.

Female jackrabbits are larger than males. They have grey fur with reddish and brown flecks. Their undersides are paler, and their tails and the tips of their huge ears are black. Like other hares, male jackrabbits indulge in frenzied fights during the breeding season.

MARSUPIALS

Marsupials are a group of mammals that brood their young in pouches on their bellies, rather than in wombs, like placental mammals. The overwhelming majority of marsupials are found in Australia and New Guinea, but several species live in the Americas. However, fossil evidence has led zoologists to believe that marsupials first evolved in South America, and spread to Australasia from there overland.

Water opossum

Chironectes minimus

Water opossums have short waterproof coats with a grey and black pattern. Their hind feet are webbed, and both sexes have pouches opening to the rear.

Water opossums, or yapoks, live beside bodies of fresh water in tropical forests. They make dens in burrows in the banks of streams or lakes, with entrances just above water level. Unusually, both sexes have pouches. A female can close her pouch using a ring of muscles to keep her developing young dry while she is underwater. A male's pouch is always open and he uses it to protect his scrotum while in water or when moving quickly through forest.

Water opossums are superb swimmers. They use their hind feet to propel themselves through the water. However, they also forage on land or in trees. They spend the night in their dens, but may rest in bundles of leaves in secluded places on the forest floor between daytime feeding forays. Most births take place between December and January. After their birth, the young spend a few more weeks in the pouch until they are fully developed.

Distribution: Central and South America from southern Mexico to Belize and Argentina.
Habitat: Freshwater streams and lakes.
Food: Crayfish, shrimp, fish, fruit and water plants.
Size: 27–40cm (10.5–16in); 600–800g (1.25–1.75lb).
Maturity: Not known.
Breeding: 2–5 young born in summer.
Life span: 3 years.
Status: Lower risk.

Virginia opossum

Didelphis virginiana

Distribution: United States, Central America and northern South America.
Habitat: Moist woodlands or thick brush in swamps.
Food: Plants, carrion, small vertebrates and invertebrates.
Size: 33–50cm (13–20in); 2–5.5kg (4.5–12lb).
Maturity: 6–8 months.
Breeding: 2 litters per year.
Life span: 3 years.
Status: Lower risk.

Virginia opossums generally live in forested areas that receive plenty of rain. However, the species is very adaptable and is making its home in new places across North America. Many survive in more open country beside streams or in swamps, while others make their homes in people's sheds and barns.

Virginia opossums are most active at night. By day they rest in nests of leaves and grass, hidden away in crevices, hollow trees and sometimes in burrows. By night, the marsupials hunt for food. They are good climbers, using their prehensile tails to cling to branches.

Virginia opossums do not hibernate, but they do put on fat as the days shorten with the approach of autumn. They rely on this fat to keep them going during the periods of harshest winter weather, when they cannot get out to feed. In the very coldest parts of their range, these marsupials sometimes suffer frostbite on their naked tails and thin ears.

Mating takes place in both late winter and spring. The young are only 1cm (0.4in) long and underdeveloped when born. Over 20 are born, but the mother can only suckle 13 at once, so the weaker babies die.

Virginia opossums have white faces, often with darker streaks. Their bodies are covered in shaggy coats of long grey and white hairs, but their tails are almost naked.

INSECTIVORES

Insectivores, or insect-eaters, belong to the Insectivora *order of mammals. The first mammals to develop their young in uteruses belonged to this group, and most insectivores still resemble these small, primitive animals. However, insectivores have evolved to live in a wide range of niches, including subterranean, terrestrial, aquatic and arboreal habitats.*

Giant mole shrew

Blarina brevicauda

Giant mole shrews live in most land habitats within their range, but they are hard to spot. They use their strong forepaws and flexible snouts to dig deep burrows in soft earth and, when on the surface, they scurry out of sight beneath mats of leaves or snow. However, they do climb into trees in search of food on occasion.

These small mammals feed at all times of the day and night. They rest in nests of grass and leaves made inside their tunnels or in nooks and crannies on the surface. Giant mole shrews will eat plant food, but they also hunt for small prey, such as snails, mice and insects. Their saliva contains a venom which paralyzes their prey. In the mating season, which takes place between spring and autumn, their territories expand so that they overlap those of members of the opposite sex.

Giant mole shrews have stout bodies with long, pointed snouts covered in sensitive whiskers. Their eyes are very small because they spend most of the time underground, and their ears are hidden under thick coats of grey hairs.

Distribution: Central Canada to south-eastern United States.
Habitat: All land habitats.
Food: Insects, small vertebrates, seeds and shoots.
Size: 12–14cm (5–5.5in); 15–30g (0.03–0.06lb).
Maturity: 6 weeks.
Breeding: Litters of 5–7 young born throughout summer.
Life span: 2 years.
Status: Common.

Star-nosed mole

Condylura cristata

Distribution: Eastern Canada to south-eastern United States.
Habitat: Muddy soil near water.
Food: Aquatic insects, fish, worms and crustaceans.
Size: 10–12cm (4–5in); 40–85g (0.08–0.19lb).
Maturity: 10 months.
Breeding: 2–7 young born in summer.
Life span: Not known.
Status: Endangered.

Star-nosed moles live in waterlogged soil. They dig networks of tunnels into the soil, which generally reach down as far as the water table. They push the mud and soil out of the entrances of the tunnels, making molehills in the process. The moles construct nests at the ends of tunnels, which are lined with dry grass.

Star-nosed moles are expert swimmers. They search for food at the bottom of streams and pools, using their sensitive snouts to feel their way and detect prey. In winter, star-nosed moles use tunnels with underwater entrances to get into ponds that are iced over. They feed in water during both the day and night, but they are only really active above ground during the hours of darkness.

Most births take place in early summer. The young already have the star of rays on their snouts. Breeding pairs of males and females may stay together throughout the winter and breed again the following year.

Star-nosed moles are named after the curious fleshy rays that radiate around each nostril. These rays are sensitive feelers used in the darkness below ground. Their fur is dark and dense, and is coated with water-repelling oils.

DOLPHINS AND WHALES

Whales and dolphins belong to the Cetacea *order of mammals, so they are called the cetaceans. They fall into two broad groups: baleen whales and toothed whales. Instead of teeth, baleen whales have horny plates lined with bristles for filter-feeding, and they include the rorquals and right whales. The toothed whales hunt down prey, and they include dolphins, porpoises, orcas (killer whales) and sperm whales.*

Amazon river dolphin

Inia geoffrensis

Distribution: Amazon and Orinoco River Basins.
Habitat: Dark, slow-moving river water.
Food: Small fish.
Size: 1.7–3m (5.5–10ft); 60–120kg (132–264lb).
Maturity: Not known.
Breeding: Single calf born between April and September.
Life span: 30 years.
Status: Vulnerable.

Amazon river dolphins live in the wide rivers of the Amazon Basin. During the rainy season, they move into flooded areas of forest and up swollen streams into lakes. They may become isolated in pools when waters recede, but most are able to survive by eating the river fish that are trapped with them.

Dolphins live in small groups. They are thought to defend the areas around them and will stay in an area as long as there is enough food. They breathe at least once a minute, through the nostrils on the tops of their heads. They dive down to the bottom of rivers to search for food, using their bristled snouts to root through mud and weeds.

Like other dolphins, Amazon river dolphins may use echolocation to find their way in the murky river waters. They sometimes feed in the same areas as giant otters. It may be that the way in which otters hunt drives fish out of the shallows towards the dolphins.

When young, Amazon river dolphins have metallic blue and grey upper bodies with silvery bellies. As they age, their upper bodies gradually turn pinkish. They have long snouts with sensitive bristles covering them.

Vaquita

Phocoena sinus

Most vaquitas have dark grey or black upper bodies, with paler undersides. Like other porpoises, vaquitas have blunt faces without the beak-like snouts of dolphins.

Vaquitas live in the upper area of the Gulf of California, near the mouth of the Colorado River. No other marine mammal has such a small range, and consequently vaquitas are extremely rare and may become extinct.

Vaquitas used to be able to swim up into the mouth of the Colorado. However, in recent years so much water has been removed from the river for irrigation and for supplying cities that the Colorado is little more than a trickle where it reaches the ocean. This has probably changed the composition of the Gulf waters, too. The vaquita population was also affected by the fishing industry in the Gulf. Fishermen drowned many vaquitas in their nets by accident, and their activities have also reduced the amount of fish available for the porpoises to eat.

Biologists know little about the lives of vaquitas. They probably spend most of their time alone, locating their prey close to the sea floor using echolocation. Births probably take place all year round.

Distribution: Gulf of California in the eastern Pacific.
Habitat: Coastal waters and mouth of the Colorado River.
Food: Fish and squid.
Size: 1.2–2m (4–6.5ft); 45–60kg (99–132lb).
Maturity: Not known.
Breeding: Not known.
Life span: Not known.
Status: Critically endangered.

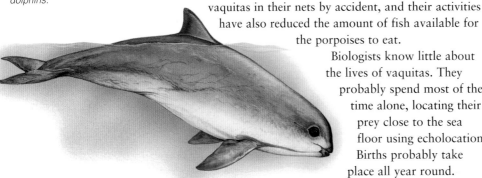

Risso's dolphin

Grampus griseus

Risso's dolphins live in small groups of about ten individuals. The groups move to warm tropical waters in winter and head back toward the poles in summer. The dolphins are often seen leaping out of the water as members of a school play with one another.

Risso's dolphins feed in deep water. They dive down to catch fast-swimming squid and fish. Like other dolphins, they probably use echolocation to locate their prey in the dark depths. They produce clicking noises that bounce off objects in the water. The dolphins can hear each other's clicks and echoes, and groups may work together to track down shoals of fish or squid. In areas where there is plenty of food, dolphin schools congregate so that thousands of the leaping mammals may be seen together.

Risso's dolphins have very blunt faces, lacking the beaks of typical dolphins. They have dark grey bodies, which are often scarred by attacks from other dolphins and large squid. Older dolphins may have so many scars that their bodies look almost white.

Distribution: All tropical and temperate seas.
Habitat: Deep ocean water.
Food: Fish and squid.
Size: 3.6–4m (12–13.25ft); 400–450kg (880–990lb).
Maturity: Not known.
Breeding: Single young born once per year.
Life span: 30 years.
Status: Common.

Spectacled porpoise (*Australophocaena dioptrica*): 1.8–2m (6–6.5ft); 115kg (253lb)
Spectacled porpoises live in coastal waters around South America and New Zealand. Their backs are black, while their undersides, including most of their faces, are white. They have black "spectacles" around their eyes. They are thought to feed on fish and squid, sometimes diving deep below the surface to catch their prey.

La Plata dolphin (*Pontoporia blainvillei*): 1.3–1.75m (4.25–5.75ft); 20–61kg (44–134lb)
These small dolphins live in the coastal waters of south-eastern South America. They are sometimes seen in the wide mouth of the Rio de la Plata in Uruguay – hence their name – but do not appear to swim up any other rivers. Their bodies are grey and they have long slender beaks that curve slightly downwards. They probe the bottom with their snouts, feeding on fish, squid and shrimp.

Tucuxi (*Sotalia fluviatilis*): 1.4–1.9m (4.5–6.25ft); 40–53kg (88–116lb)
These dolphins live in coastal waters from the Caribbean to Brazil and in the rivers of the Amazon Basin, right up to the foothills of the Andes Mountains on the western side of South America. They have grey upper bodies and pinkish undersides. Tucuxis feed mainly on fish and shrimp.

Beluga

Delphinapterus leucas

Beluga means white in Russian, so these whales are sometimes called white whales. However, they should not be confused with white sturgeon – large fish that produce beluga caviar. Belugas also have a nickname – sea canaries – because they call to each other with high-pitched trills.

Belugas live in the far north, where days are almost absent or very short for much of the year. Some beluga pods (schools or groups) spend all their time in one area of ocean, such as the Gulf of St Lawrence. Other pods are always on the move. The pods are ruled by large males, and all pods spend their winters away from areas of thick ice, which may mean being farther or nearer to land depending on where they are. In summer, they enter river estuaries and shallow bays.

They navigate using a well-developed sonar system, thought to be controlled by their melons, which are large sensory organs on top of their heads. Most calves are born in late summer, and their mothers will mate again in early summer a year or two later.

Adult beluga whales are almost completely white, helping them to hide among ice floes. Younger whales begin life with dark bodies, which gradually become yellow and brown before fading to white.

Distribution: From Arctic Ocean to Gulf of Alaska and Gulf of St Lawrence.
Habitat: Deep coastal waters and mouths of large rivers.
Food: Fish, squid, octopuses, crabs and snails.
Size: 3.4–4.6m (11.25–15.25ft); 1.3–1.5 tonnes (2860–3300lb).
Maturity: Females 5 years; males 8 years.
Breeding: Single calf born every 2–3 years.
Life span: 25 years.
Status: Vulnerable.

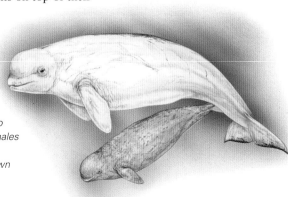

Orca

Orcinus orca

Distribution: Throughout the world's oceans.
Habitat: Coastal waters.
Food: Seals, other dolphins, fish, squid, penguins and crustaceans.
Size: 8.5–9.8m (28–32.25ft); 5.5–9 tonnes (12,000–20,000lb).
Maturity: Females 6 years; males 12 years.
Breeding: Single young born generally in autumn every 3–4 years.
Life span: 60–90 years.
Status: Lower risk.

Orcas have black upper bodies and white undersides. They also have grey patches behind their dorsal fins and white patches along their sides and above the eyes. These "whales" are really the largest members of the dolphin family.

Orcas are also known as killer whales. They are expert hunters, being armed with up to 50 large, pointed teeth, and they catch prey in all areas of the ocean. Although orcas have been detected 1km (0.6 mile) below the surface, they prefer to hunt in shallow coastal waters and often swim into bays and mouths of rivers to snatch food near the shore.

Orcas typically live in pods of five or six individuals. Generally each pod is run by a large male, although larger groups have several adult males. Females and their young may split off into subgroups. Like other toothed whales and dolphins, orcas produce click sounds that are used for echolocation. The whales also communicate with each other using high-pitched screams and whistles. Orcas have several hunting techniques. They break pack ice from beneath, knocking their prey into the water, or they may rush into shallow water to grab prey from the shore. It is reported that they may crash on to the shore to drive prey into the surf where other members of the pod pick them off. Orcas breed throughout the year, although most mate in the early summer and give birth in the autumn of the following year.

Humpback whale

Megaptera novaeangliae

Humpbacks spend their summers feeding far from shore, in the cold waters near the poles. They feed by taking in huge mouthfuls of sea water. Their baleen plates then strain out any fish or krill from the water. Pairs of humpbacks also corral schools of fish by blowing curtains of bubbles around them. The fish will not swim through the bubbles and crowd together as the whales rush up from beneath with their mouths wide open.

As winter approaches, the whales stop feeding and head to warmer, shallow waters near coasts or groups of islands. For example, populations of humpbacks spend the winter near Baja California and the Hawaiian islands. During the winter the whales do not feed; instead they concentrate on reproduction. The males produce songs which are repeated over and over for days on end. The songs probably attract receptive females that are not caring for calves that year, and also help rival males keep away from each other. Pregnant females stay feeding for longer than the other whales, and arrive in the wintering grounds just in time to give birth.

Humpback whales are so called because of their dorsal fins, which may be swelled into humps by deposits of fat. Humpbacks have the longest pectoral (arm) fins of any whale – about a third as long as their bodies. These baleen whales have throat grooves which expand to enlarge the throat size as the feeding whale gulps water.

Distribution: All oceans.
Habitat: Deep ocean water.
Food: Small fish and krill.
Size: 12.5–15m (41–49.5ft); 30 tonnes (66,000lb).
Maturity: 4–5 years.
Breeding: Single young born every 2 years.
Life span: 70 years.
Status: Vulnerable.

Grey whale

Eschrichtius robustus

Grey whales spend their lives on the move. In autumn they swim from Arctic waters down the western coast of North America, mating on the way, to spend the winter in bays along the coast of Mexico. The young that were conceived during the previous year are born in these bays in late January and February, and soon after, the whales set off to spend the summer in the food-rich waters of the Arctic.

A similar migration takes place down the eastern coast of Asia, but these whales are relatively few in number. Grey whales spend a good deal of time playing in shallow water during the winter. They leap out of the water and may become stranded for a few hours as they wait for the tide to rise. While on the move, they "spyhop" – protrude their heads above the surface so that they can look around.

Grey whales are baleen whales that feed on the seabed. They drive their heads through the sediment to stir up prey. They then suck in the disturbed water and strain the animals from it. Most feeding takes place in the summer, and whales may fast for the remaining six months of the year.

Distribution: Northern Pacific Rim.
Habitat: Shallow coastal water.
Food: Amphipods (small crustaceans).
Size: 13–15m (43–49.5ft); 20–37 tonnes (44,000–81,000lb).
Maturity: Females 17 years; males 19 years.
Breeding: Single young born every 2 years.
Life span: 70 years.
Status: Endangered

Grey whales do not have dorsal fins, but a series of small humps along their backs. They are often covered in white barnacles.

Bowhead whale (*Balaena mysticetus*): 11–13m (36.25–43ft); 50–60 tonnes (110,000–132,000lb)
Bowhead whales live in the Arctic Ocean. They have huge curved jaws with more baleen plates than any other whale. Adults have black bodies with pale patches on their lower jaws. Bowheads live among ice floes. They feed on tiny floating crustaceans, such as krill and copepods. They can eat 1.8 tonnes (3,960lb) in one day.

Blue whale (*Balaenoptera musculus*): 25–30m (82.5–100ft); 100–160 tonnes (220,000–352,000lb)
The blue whale is the largest animal to have ever existed. It has a blue-grey body, with spots along its back and a pale pleated throat. Blue whales live alone, ploughing between subtropical waters and those near the poles. The populations of the northern and southern hemispheres never meet. They eat krill – tiny floating crustaceans – and can gulp down 6 tonnes (13,200lb) of them in a single day. Their pleated throats distend to four times their normal size as the whales take in mouthfuls of krill-laden water.

West Indian manatee

Trichechus manatus

Despite appearances, manatees are not cetaceans. Neither are they related to seals – pinnipeds. These marine mammals belong to the *Sirenia* order, as do the dugongs – similar animals from South-east Asia. Sirenians evolved to live in water separately from whales and seals. In fact, they are believed to be more closely related to elephants than other sea mammals. Like elephants, they are vegetarian, not carnivorous.

Manatees live in both salt and fresh water, although they spend more time in freshwater habitats. They rarely stray far from land, and may travel far up rivers to sources of warm water during winter.

Manatees feed both during the day and night. They use their dextrous lips to pluck water plants, such as water hyacinths. Although they do not actively seek them out, manatees' vegetarian diets are supplemented by the aquatic invertebrates, such as snails and insect larvae, living on the water plants. The single young is born after about a year's gestation.

Distribution: Coast of Florida to Brazil.
Habitat: Estuaries and shallow coastal water.
Food: Water plants and aquatic invertebrates.
Size: 2.5–4.5m (8.25–15ft); 500kg (1100lb).
Maturity: 8–10 years.
Breeding: Single young born every 2–3 years.
Life span: 30 years.
Status: Vulnerable.

Manatees have wrinkled grey-brown skin with a sparse covering of fine hairs. Their fore-flippers have nails on their upper surfaces and their upper lips, which are very manoeuvrable, have moustaches of thick bristles.

SEALS AND RELATIVES

Seals, sea lions and walruses are pinnipeds – they belong to the Pinnipedia *order of mammals. They are descended from carnivorous, terrestrial ancestors. However, it seems that seals may be only distantly related to sea lions and walruses, despite their similar appearances. Like other sea mammals, pinnipeds have a thick layer of blubber under their skin, which keeps them warm while swimming in cold water.*

Californian sea lion

Zalophus californianus

Californian sea lions spend the year moving up and down the Pacific coast of North America. In autumn and winter, most males move north to feed off the coast of British Columbia. The females and young stray less far from the breeding grounds and probably head south at this time.

The sea lions are seldom far from the shore at any time of the year. They generally go on foraging trips at night, although they are often active during the day as well. Each trip can last for several hours.

During the summer breeding season, the sea lions congregate on flat beaches in the central area of their range. Most choose sandy habitats, but will use open, rocky areas if necessary. The males arrive first and fight each other for control of small territories on the beaches and in the water. They can only hold their territories for a few weeks before having to swim away and feed.

Californian sea lions have less heavyset bodies than most sea lions because they live in warmer climes. They are fast swimmers, reaching speeds of 40kph (25mph).

Distribution: Pacific coast of North America and Galápagos Islands.
Habitat: Ocean islands and coastline.
Food: Fish, squid and seabirds.
Size: 1.5–2.5m (5–8.25ft); 200–400kg (440–880lb).
Maturity: Females 6 years; males 9 years.
Breeding: Single pup born each year.
Life span: 20 years.
Status: Vulnerable.

South American sea lion

Otaria flavescens or *byronia*

South American sea lions do not travel very far from their breeding sites during the non-breeding season, although they may spend long periods out at sea. These sea lions sometimes feed in groups, especially when they are hunting shoals of fish or squid.

The breeding season begins at the start of the southern summer. Adults arrive on beaches or flat areas of rock at the beginning of December. Males arrive a few weeks before the females, and defend small patches of the beach. The females give birth to the young they have been carrying since the year before. After a few weeks of nursing their pups, the females become receptive to mating again. As the numbers of females increases, males stop controlling territories and begin to defend groups of females. Unsuccessful males without harems of their own gang together on the fringes of the beaches and charge through the females to mate with them.

Male South American sea lions have dark brown bodies with brown manes on their heads and necks. The females are less heavyset and have paler bodies and no manes.

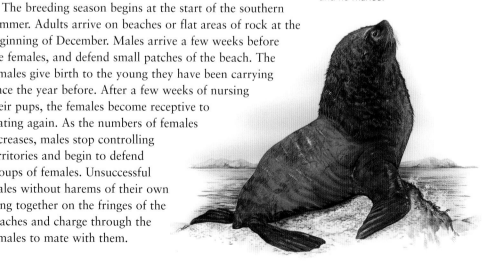

Distribution: South Pacific and Atlantic waters off the South American coast from northern Peru to Brazil.
Habitat: Coastal waters and beaches.
Food: Fish, squid and crustaceans.
Size: 1.8–2.5m (6–8.25ft); 150–350kg (330–770lb).
Maturity: Females 4 years; males 6 years.
Breeding: Single pup born in January.
Life span: 20 years.
Status: Vulnerable

Walrus

Odobenus rosmarus

Walruses live among the ice floes of the Arctic Ocean. These huge sea mammals are well known for their long tusks, which they use to stab opponents during fights. Walruses also use their tusks to "haul out", or pull themselves on to floating ice, and sometimes hook themselves to floes so that they can sleep while still in the water.

Walruses use their whiskered snouts to root out prey and blast away sediment with jets of air squirted from the mouth. They tackle shelled prey by holding them in their lips and sucking out the soft bodies.

Walruses live in large herds, sometimes of many thousands. In winter they feed in areas of thin sea ice, avoiding thick, unbroken ice, which they cannot break through from beneath. In summer, when the ice recedes, they spend more time on land.

Walruses have long tusks growing out of their upper jaws. Males, which are twice the size of females, also have longer tusks. Their bodies are reddish-brown and sparsely covered in coarse hairs. Males have two air pouches inside their necks, which they use to amplify their mating calls.

Distribution: Coast of Arctic Ocean.
Habitat: Pack ice.
Food: Worms, shellfish and fish.
Size: 2.25–3.5m (7.5–11.5ft); 400–1,700kg (880–3,740lb).
Maturity: Females 6 years; males 10 years.
Breeding: Single young born once per year.
Life span: 40 years.
Status: Vulnerable.

Southern elephant seal (*Mirounga leonina*): 2–9m (6.5–30ft); 400–5,000kg (880–11,000lb)
Southern elephant seals are the largest pinnipeds in the world. The males are four to five times the size of the females and have inflatable trunk-like noses, with which they amplify their bellowing calls. Elephant seals live in the waters around Antarctica and South America, where they dive for fish and squid. They congregate on flat beaches to breed in winter. The males arrive first and take up positions. They may stay on the beaches for the next two months without feeding. The females arrive a few weeks later and give birth. After giving birth, the females spread out and the males battle to mate with harems of females along the beaches. On a smaller beach, all the females may be controlled by a single bull, known as a beach master.

Steller's sea lion (*Eumetopias jubatus*): 2.4–3m (8–10ft); 270–1,000kg (594–2,200lb)
Steller's sea lions are the largest sea lions. They live along the coast of the northern Pacific Ocean, from Japan to California. Adult males have huge necks, made even chunkier by coarse manes. The females are about a third of their size. Steller's sea lions often rest on rocky coasts, but they feed on fish, squid and shellfish out at sea. Females give birth to single pups in early summer.

Hooded seal

Cystophora cristata

Hooded seals rarely approach land, preferring to spend their whole lives among the ice floes in the cold Arctic Ocean. Apart from during the breeding season, hooded seals live alone. They dive down to depths of more than 180m (590ft) to catch shoaling fish and bottom-living creatures.

When the breeding season arrives in spring, the seals congregate on wide ice floes. The females take up widely spaced positions on the ice, preparing to give birth to the young conceived the year before. Meanwhile, males compete for access to small groups of females. The victors stay near the females as they nurse their new-born calves, chasing away any intruders while inflating their nasal balloons.

Hooded seal pups are suckled for only four days – the shortest time of any mammal – after which the mothers abandon them.

Distribution: Waters around Greenland.
Habitat: Drifting ice floes.
Food: Octopuses, squid, shrimp, mussels and fish.
Size: 2–2.7m (6.5–9ft); 145–300kg (320–660lb).
Maturity: Females 3 years; males 5 years.
Breeding: Single pup born in March.
Life span: 30 years.
Status: Common.

Hooded seals are so named because the males possess elastic sacs or hoods on the tops of their heads. The hoods are connected to their noses and can be inflated with air to amplify their calls while sparring with rival males. Female seals also have hoods, but they are not inflatable.

INDEX

ACKNOWLEDGEMENTS
The publisher would like
to thank the following for
permission to reproduce their
photographs in this book.

Key: l=left, r=right, t=top,
m=middle, b=bottom

NHPA: 8bl, 16tr, 16b, 18t, 18b,
19t, 21t, 21br, 24t, 26t, 27tl, 29b,
30t, 31t, 31b.
Tim Ellerby: 23tl, 28l, 31m.